A HISTORY OF GENETICS

A HISTORY OF GENETICS

A. H. STURTEVANT

Thomas Hunt Morgan
Professor of Biology, Emeritus

California Institute of Technology

WITH AN INTRODUCTION BY EDWARD B. LEWIS

Cold Spring Harbor Laboratory Press
Cold Spring Harbor, New York

ESP   Electronic Scholarly Publishing Project
http://www.esp.org

A History of Genetics

© 1965, 1967 Alfred H. Sturtevant
© 2001 (Note from the Publishers and Introduction) Cold Spring Harbor Laboratory
Press and Electronic Scholarly Publishing Project
Published by Cold Spring Harbor Laboratory Press and Electronic Scholarly Publishing Project

Electronic Scholarly Publishing Project
ESP Foundations Reprint Series: Classical Genetics
Series Editor: Robert J. Robbins
Electronic Production: Ana Dos Santos, William Landram

Cold Spring Harbor Laboratory Press
Acquisition Editor: John R. Inglis
Production Editor: Mala Mazzullo
Cover Designers: Robert J. Robbins / Danny deBruin

Front Cover: Species of Drosophila, by Edith M. Wallace, Plate 3, from *The North American Species of* Drosophila, by A. H. Sturtevant, 1921, Carnegie Institution of Washington.

Back Cover: Sex-linked inheritance, Figure 38 from *The Theory of the Gene*, by T. H. Morgan, 1926, Yale University Press.

Bibliographical Note: This CSHLP/ESP edition, first published in 2000, is a newly typeset, unabridged version of *A History of Genetics*, by A. H. Sturtevant, based on the second printing of the original 1965 book, published by Harper & Row in their Modern Perspectives in Biology series. All footnotes were taken from the original work. The figures have been redrawn for this edition.

Availability: A directory of Cold Spring Harbor Laboratory Press publications is available at http://www.cshl.org/books/directory.htm. All Electronic Scholarly Publishing Project publications may be obtained online, from http://www.esp.org.

Web site: The full text of this book, along with other supporting materials, including full-text copies of many of the works discussed by the author, may be obtained at the book's web site: http://www.esp.org/books/sturt/history.

Library of Congress Cataloging-in-Publication Data
Sturtevant, A. H. (Alfred Henry), 1891-
    A history of genetics / A. H. Sturtevant
        p. cm.
    Previously published: New York: Harper & Row, 1965.
    Includes bibliographical references and index.
    ISBN 0-87969-607-9 (pbk.: alk. paper)
        1. Genetics--History. I. Title
QH428 .S78 2000
576.5'09--dc21   00-052394

10 9 8 7 6 5 4 3

Text reprinted with permission from the family of A. H. Sturtevant.

CONTENTS

NOTE FROM THE PUBLISHERS

This book is special in several rather diverse ways. First published by Alfred Sturtevant in 1965, it is one of the very few accounts of the early days of genetics by one who was there—the truths of a reporter rather than an historian. Sturtevant was one of an accomplished trio of Thomas Hunt Morgan's students, and although his name may resonate less with today's scientists than the names of his colleagues Bridges and Muller, his keen intelligence and broad scientific interests gave his book a scope of unusual breadth and interest. Yet it did not endure. A second printing appeared in 1967. Three years later Sturtevant was dead, and increasingly rare copies of his book were consigned to library shelves and second-hand shops as the concepts and techniques of molecular biology swept to dominance in the field of genetics.

This reprinted edition has its origins in two independent initiatives. Prompted by colleagues on the scientific staff, Cold Spring Harbor Laboratory Press has in recent years republished two long-out-of-print books with both historical interest and continued contemporary relevance: *The Biology of* Drosophila by Milislav Demerec and *The Structure and Reproduction of Corn* by Theodore Kiesselbach. The response to these volumes was warm and encouraging, so when the idea of reviving Sturtevant's classic was suggested, we were enthusiastic, particularly when it was pointed out that Sturtevant's student and recent Nobel Prize winner, Edward Lewis, might be persuaded to write a new introduction to the book. Dr. Lewis kindly agreed to the task and did his part quickly and well. However, the currently rapid rate of growth and expansion within the Press meant that momentum on the project slowed, since the project lacked the urgency of books with the latest research results that are our typical output.

Independently, Robert Robbins, a biologist turned information scientist with a long-standing interest in both the history of science and the technology of publishing, had become interested in seeing the book return to print. Intrigued by the possibilities of networked information, he had established the Electronic Scholarly Publishing Project, a web-based repository of historically interesting books and papers displayed in a way that leveraged the unique advantages of online delivery—full text-based searching, links to other electronic information sources, and personal annotation of the stored document. The ESP Project places a special emphasis on works related to the foundations of classical genetics.

Robbins' desire to add the Sturtevant book to this repository led him first to the Sturtevant family, then to Ed Lewis, then to Cold Spring Harbor Laboratory itself, with the result that the Electronic Scholarly Publishing Project and the Cold Spring Harbor Laboratory Press agreed to produce the book jointly, with an online and a print version to appear simultaneously.

The outcome is the book you hold in your hands. Along with the physical book, we have also produced a website associated with the project. At that site, readers may obtain full-text electronic versions of many of the key papers discussed by Sturtevant, including Sturtevant's own "The linear arrangement of six sex-linked factors in Drosophila, as shown by their mode of association," which contained the world's first genetic map. The book's website can be seen at http://www.esp.org/books/sturt/history.

The partnership of Cold Spring Harbor Laboratory Press with the Electronic Scholarly Publishing Project is an experiment, one of many being conducted in this era of new publishing paradigms. It is our hope that for the reader, this print–online combination will deliver the best of both media, as a vehicle of an exceptional work of scholarship that deserves fresh recognition by a new generation of scientists.

We are pleased that this book appears in the year 2000—a year with special significance for genetics and for the study of *Drosophila melanogaster*. This is the 100th anniversary of the founding of modern genetics with the rediscovery of Mendel's work, and it is the year in which the full DNA sequence of the Drosophila genome was obtained. The fruit fly is still at the center of genetic research, just as it was when Sturtevant first began his work in the "Fly Room" at Columbia University.

JOHN R. INGLIS
Cold Spring Harbor Laboratory Press

ROBERT J. ROBBINS
Electronic Scholarly Publishing Project

INTRODUCTION

The reprinting of this classic book provides students with one of the few authoritative, analytical works dealing with the early history of genetics. Those of us who had the privilege of knowing and working with Sturtevant benefited greatly from hearing first-hand his accounts of that history as he knew it and, in many instances, experienced it. Fortunately, Sturtevant put it all together in this book.

In his preface to the book, Sturtevant lists the persons that he knew personally and who were major players in the field, in addition to those who occupied the famous fly room (Chapter 7) at Columbia University. As a result, much of the history is based on first-hand contacts as well as on a scholarly and critical review of the literature of genetics and cytology.

Sturtevant was clearly present at the creation of modern genetics, if dated from 1910 when Morgan commenced work on Drosophila. Of Morgan's three students—Sturtevant, Bridges, and Muller—Sturtevant was ideally suited to write the history because of his remarkable memory, his knowledge of almost all aspects of biology, and a keen analytical ability that extended not only to his experimental work, but also to tracing the history of the underlying ideas.

Sturtevant was a gifted writer and also an authority on many of the subjects he covers. While he was a sophomore in college, he deduced the linear order of the genes. Later, he postulated the existence of inversions and duplications before they were verified cytologically. Sturtevant was especially interested in how genes produce their effects and, consequently, was the father of a field now called developmental genetics. In this area, his style was to analyze exceptions to the rule. In so doing, he identified the phenomenon of position effect, in which the position of a gene (that of the Bar, and double-Bar, eye mutations) can be shown to affect its function. He identified the first clear case of a non-autonomously expressed gene, vermilion, mutants of which produce a vermilion, instead of the normal red, eye color. This was an important exception to the rule that sex-linked mutants behaved autonomously in gynandromorphs. How this led to the field of biochemical genetics is explained in Chapter 16.

In the tradition of such biologists as Darwin, Galton, and Bateson and of many of the early Mendelians, Sturtevant was an ardent evolutionist. He had a seemingly inexhaustible knowledge of embryology, anatomy, morphology, and taxonomy that served him well in treating evolutionary concepts historically, as described in Chapter 17. It is a wide-ranging chapter that covers many topics, including the development of population genetics, the role of gene mutations in evolution, and, prophetically, the conservation of bio-chemical pathways in major groups from bacteria to vertebrates. His own experimental work, typically only briefly referred to, included his work on interracial and interspecific hybrids in the genus Drosophila, and the demonstration that the genetic content of different species of that genus is remarkably conserved, whether it be in the X chromosome or in each of the specific autosomal arms. Sturtevant always had a healthy skepticism, surely one of the most important qualities of a successful scientist. This is shown by his doubt of the value of many laboratory experiments in population genetics, on the basis that they cannot faithfully duplicate what really goes on in the great out-of-doors.

It may come as a surprise to many students to realize how much opposition there was in some quarters to the early discoveries of the Morgan school. Sturtevant's account of such controversies is a recurrent theme of this book, as it should be in a historical treatise.

Science has often been advanced by scientists who questioned existing dogma and found it flawed. Or, conversely, such dogma has probably in some cases slowed progress for years. Would advances in genetics have been more rapid had there not been the virtually universal belief that genes were proteins, or that development of an organism involved cytoplasmic rather than nuclear heredity? Sturtevant does not waste space speculating about such issues, but he does discuss several cases in which progress was held back because of failure to develop a satisfactory terminology and symbolism.

Sturtevant had a strong social consciousness that comes forth in Chapter 20. There he treats the history of human genetics, stressing the difficulties and pitfalls that plague studies in this field. He devotes considerable space to an objective and critical analysis of the so-called "nature vs. nurture" question.

In the last chapter, Sturtevant discusses how discoveries in science and particularly genetics tend to come about. He addresses in his typically analytical way the often-cited dictum: The time has to be ripe for a discovery to be made and that when that time comes someone is bound to make the discovery. He concludes that this attitude greatly oversimplifies what generally happens in science.

I believe Sturtevant's writing of this book after his retirement was one more intellectual exercise to stave off boredom. He had reduced his experimental work to an hour or so each day, and it must have been more difficult to keep up with the expanding literature of the field. His book is clearly a labor of love and his personality shines through every page.

July 2000   E. B. Lewis
Pasadena, California

AUTHOR'S PREFACE

The publication of Mendel's paper of 1866 is the outstanding event in the history of genetics; but, as is well known, the paper was overlooked until 1900, when it was found. Its importance was then at once widely recognized. These facts make the selection of topics for the early chapters of this book almost automatic. What was the state of knowledge about heredity in Mendel's day; what sort of man was Mendel, and how did he come to make his discovery; what happened between 1866 and 1900 to account for the different reaction to his results; how did his paper come to be found, and just what was the immediate reaction?

These questions are discussed in the first four chapters. From that point on, it has seemed more logical to treat the various topics separately rather than to follow a more nearly chronological order. The attempt has been, in each case, to trace the beginnings of a subject and to bring it down to the work currently being done—but not to discuss presently active fields of work, since these are adequately covered in current books and reviews. There is no definite terminal date, but work later than about 1950 is generally omitted or is referred to only briefly. In other fields the cutoff date is even earlier than this.

For Chapters 1 and 3 I have relied largely on secondary sources such as Sachs (1875), Zirkle (1935), Roberts (1929), and Wilson (1925). For the period after 1900 I have read or reread much of the original literature and, for general background, have been fortunate enough to have had some direct personal contact with many of the people discussed—including, among the early workers, de Vries, Bateson, Johannsen, Wilson, Morgan, McClung, East, Shull, Castle, Emerson, Davenport, Punnett, Nilsson-Ehle, Goldschmidt, and others. (I have seen Cuénot, Baur, Sutton, and Saunders but never really knew them.)

I am indebted to numerous colleagues who have read part or all of the manuscript and have made constructive suggestions. Especially to be named are Drs. N. H. Horowitz, E. B. Lewis, H. L. Roman, C. Stern, G. Hardin, and C. Fulton. Much of the material has been presented in a series of lectures at the California Institute of Technology and at the Universities of Washington, Texas, and Wisconsin; numerous discussions with colleagues at these institutions have been very helpful.

August 1965   A. H. STURTEVANT
Pasadena, California

xi

CHAPTER 1

BEFORE MENDEL

In discussing the history of a subject it is usual to begin with Aristotle—and he forms a convenient starting point for genetics, though the real beginnings, even of theoretical genetics, go farther back. As a matter of fact, much of Aristotle's discussion of the subject is contained in his criticism of the earlier views of Hippocrates.

Hippocrates had developed a theory resembling that later proposed by Darwin, who called it "pangenesis." According to this view, each part of the body produces something (called "gemmules" by Darwin) which is then somehow collected in the "semen"—or as we should now say, the germ cells. These are the material basis of heredity, since they develop into the characters of the offspring. The view was developed, both by Hippocrates and by Darwin, largely to explain the supposed inheritance of acquired characters. Aristotle devoted a long passage to criticism of this hypothesis, which he discarded for several reasons. He pointed out that individuals sometimes resemble remote ancestors rather than their immediate parents (which is in fact one of the arguments used by Darwin *for*, rather than *against*, pangenesis, since Darwin did not suppose that the gemmules necessarily came to expression in the first generation and did not suppose, as did Hippocrates, that they were released from the parts of the body at the moment of copulation). Aristotle also pointed out that peculiarities of hair and nails, and even of gait and other habits of movement, may reappear in offspring, and that these are difficult to interpret in terms of a simple form of the hypothesis. Characters not yet present in an individual may also be inherited—such things as gray hair or type of beard from a young father—even before his beard or grayness develops. More important, he pointed out that the effects of mutilations or loss of parts, both in animals and in plants, are often not inherited. Aristotle, like everyone else until much later, accepted the inheritance of acquired characters; but he was nevertheless aware that there was no simple one-to-one relation between the presence of a part in parents and

1

its development in their offspring. His general conclusion was that what is inherited is not characters themselves in any sense but only the potentiality of producing them. Today this sounds self-evident, but at that time it was an important conclusion, which was not always fully understood, even by the early Mendelians.

Aristotle was a naturalist and described many kinds of animals—some imaginary, others real and described in surprisingly accurate detail. He knew about the mule and supposed that other animals were species hybrids—that the giraffe, for example, was a hybrid between the camel and the leopard. According to him, in the dry country of Libya there are few places where water is available; therefore many kinds of animals congregate around the water holes. If they are somewhere near the same size, and have similar gestation periods, they may cross; this is the basis for the saying that "something new is always coming from Libya."

Some later authorities disregarded Aristotle's reasonable limitations on what forms might be expected to cross, as in the conclusion that the ostrich is a hybrid between the sparrow and the camel. There is a long history of such supposed hybrids—notably of the crossing between the viper and the eel, and of the hybrid between the horse and the cow. Zirkle records accounts of both of these as late as the seventeenth century.

The knowledge of sex in animals goes far back before the beginnings of history and was understood quite early even in plants—at least in two important food plants of the Near East, namely, the Smyrna fig and the date palm, both of which are dioecious (that is to say, have separate male and female trees). Zirkle shows that a special Near Eastern deity (the cherub) was supposed to preside over the date pollination, and that representations of this deity can be traced back to about 1000 B.C. There is, in fact, evidence that male and female trees were grown separately as early as 2400 B.C.

The condition found in these two trees was definitely related to the phenomenon of sex in animals, by Aristotle and others, but it was much later that it was realized that plants in general have a sexual process.

That the higher plants do have sexual reproduction and that the pollen represents the male element seems to have been first indicated as an important generalization by Nehemiah Grew in 1676. A sound experimental basis was first given by Camerarius (1691 to 1694). From that time on, the view was rather generally accepted, especially after Linnaeus presented more evidence and lent the prestige of his name in 1760.

More or less casual observations on natural or accidental hybrids in plants were made over a long period, beginning with the observations of Cotton Mather on maize in 1716. However, the systematic study of plant hybrids dates from the work of Kölreuter, published from 1761 to 1766. His work laid the foundations of the subject and was familiar to Darwin and to Mendel, both of whom discussed it a hundred years later.

Kölreuter made many crosses, studied the pollination process itself, and also recognized the importance of insects in natural pollination. He used a simple microscope to study the structure of pollen and was the first to describe the diversity of pollen grains found in seed plants. He also made studies on the germination of pollen. These studies on germination were carried out on pollen in water, with the result that the pollen tubes plasmolyzed almost immediately. This led Kölreuter to conclude that the fertilizing agent was the fluid released on the stigma, rather than a formed element from a particular pollen grain.

In another respect he reached a wrong conclusion that delayed the development of a clear understanding of fertilization, namely, the view that more than one pollen grain is necessary for the production of a normal seed. This view was based on experiments with counted numbers of pollen grains, which seemed conclusive to him. The result was generally accepted for some time, and even Darwin adopted it (*The Variation of Animals and Plants under Domestication,* Ch. 27) on the authority of Kölreuter, and of Gärtner, who later confirmed the experiments. Kölreuter supposed, as a result of his experiments, that he could recognize "half-hybrids," that is, individual plants derived from pollen that was partly from the seed parent and partly from a different plant. Like Aristotle and other predecessors, he thought of fertilization as resulting from a mixing of fluids, basing this in part on his direct observations of germinating pollen.

His observations on the hybrids themselves were of importance. He recognized that they were usually intermediate between the parents (he was nearly always using strains that differed in many respects), but he did record a few cases where they resembled one parent. He recognized the sterility often found in hybrids between widely different forms and showed that in some of these the pollen was empty. He emphasized the identity of the hybrids from reciprocal crosses—which is rather surprising, since plastid differences might have been expected in some of such a large number of reciprocal species hybrids.

Kölreuter reported a few instances of increased variability in the offspring of hybrids but laid no emphasis on this observation. He also observed the frequent great increase in the vegetative vigor of hybrids and suggested that it might be of economic importance, especially if hybrid timber trees could be produced.

Following Kölreuter, there were a number of men engaged in the study of plant hybrids. Detailed accounts of their work are given by Roberts (1929), but perhaps the most satisfactory general account of the state of knowledge in Mendel's time is to be found in Darwin's discussion in *The Variation in Animals and Plants under Domestication* (1868).*

Darwin collected a vast amount of information from the works of the plant hybridizers, from works on the practical breeding of domestic animals and cultivated plants, and from gardeners, sportsmen, and fanciers. He himself carried out numerous experiments with pigeons and with various plants. The book is still interesting, as a source of information and of curious observations. Darwin was looking for generalizations, and extracting them from masses of observations was his special ability. But, in the case of heredity, the method yielded very little. He recognized two more or less distinct types of variations—those that came to be known as continuous and discontinuous, respectively. The latter, sometimes called "sports," he recognized as sometimes showing dominance, and as being often transmitted unchanged through many generations. But he felt that they were relatively unimportant as compared to the continuously varying characters, which could be changed gradually by selection and which gave intermediate hybrids on crossing. He concluded that crossing has a unifying effect. Since hybrids are generally intermediate between their parents, crossing tends to keep populations uniform, while inbreeding tends to lead to differences between populations; this same conclusion is shared by modern genetics, though the arguments are not quite the same as Darwin's.

He reported crosses which led to increased variability in the second and later generations, but he was interested in them chiefly because of their bearing on the question of reversion to ancestral types. He also recognized the increase in vigor that often results from crossing and observed the usual decline due to continued inbreeding. He carried out numerous detailed experiments in this field, which are elaborated in one of his later books, *(The Effects of Cross and Self Fertilization in the Vegetable Kingdom,* 1876).

On the origin of variability, Darwin had little to say that sounds

---

* Darwin's books were extensively altered in successive editions, and it is not always safe to consult a later edition and then to assign the views given therein to the date of the first edition. Although I have not seen the first edition of the book, I have no reason to suppose that its date is misleading in this connection.

modern. He thought that changed conditions, such as domestication, stimulated variability and also affected the inheritance both in selection within a strain and in crosses between strains. The effects of selection were familiar to him, but he was not aware of the basic distinction between genetically and environmentally produced small variations.

Darwin's own theory of heredity (pangenesis) was not generally well received, but it did apparently serve to suggest the particulate theories of Weismann and of de Vries, which paved the way in 1900 for the appreciation of Mendel's work.

The development of ideas about inheritance in animals and in plants was rather independent, for in plants the early experiments were directed largely toward the demonstration of sexual reproduction, which needed no demonstration in animals. This led to the study of hybrid plants, but in animals the development was largely in the hands of practical breeders, who were more concerned with selection than with crossing. One of the striking things about Darwin was that he had a detailed firsthand knowledge of both animals and plants, and of the literature on both. In his work we find the modern custom of discussing theory without regard to the distinction between animals and plants. It is true that this had been done before—by Aristotle, for example—but not to the extent that Darwin introduced. It may be noted that the previous hybridizers referred to in Mendel's paper (Kölreuter, Gärtner, Herbert, Lecoq, and Wichura) were all botanists. Since Mendel referred to them, we may suppose that they influenced his work; therefore there follow brief accounts of the last four, since Kölreuter has already been discussed.

Gärtner's work was published largely in 1839 and in 1849. He made a large number of crosses. Roberts says that "he carried out nearly 10,000 separate experiments in crossing, among 700 species, belonging to 80 different genera of plants, and obtained in all some 350 different hybrid plants." In general, he confirmed much of Kölreuter's work, but added little that was new, except for an insistence on the greater variability of F_2 (the second generation) compared to F_1 (the first generation). He did not often describe the separate characters of his plants but rather treated them as whole organisms—a habit common to many of the older hybridizers. Mendel gave a good deal of space to a discussion of Gärtner's results. He interpreted them as due in part to the multiplicity of gene differences between the plants crossed—which in F_2 resulted in great rarity of individuals closely resembling the parents. Gärtner also carried out experiments with several plants that involved back-crossing hybrids in successive generations to one of the parental species, in an effort to see how many such backcrosses would be needed to eliminate the characters of the other parent. Mendel did a few experiments of this kind with peas and found, as he expected, that the result depends on the proportion of dominant genes in the parent to which the back-crossing is done. He suggested that this factor must always complicate experiments of the kind carried out by Gärtner (and earlier similar crosses made by Kölreuter).

The work of Herbert, published between 1819 and 1847, dealt chiefly with crosses among ornamental plants. Perhaps his most important contribution was his discussion of the idea that crosses between species are unsuccessful or yield sterile hybrids, while crosses between varieties yield fertile offspring. He pointed out that there is no sharp line here, and that the degree of structural difference between two forms is not an invariable index of the fertility of their hybrids. In short, the argument is a circular one: infertility between species and fertility between varieties can be concluded only if fertility and sterility are made the criteria by which species and varieties are defined.

Lecoq (published 1827 to 1862) was interested in the breeding of improved agricultural plants. He made many crosses and discussed the results of other hybridizers, but seems to have added little that advanced the subject.

Wichura's chief paper appeared in 1865, after Mendel's experiments were completed; he therefore could scarcely have influenced the planning of Mendel's crosses. His work was on the crossing of willows; perhaps the most striking passages have to do with the necessity for extreme care in preventing unwanted pollen from confusing the experiments, and his strong insistence on the identity of reciprocal hybrids—the latter being a point that Lecoq had believed was not correct.

Two other people in this period should be discussed, since both have been cited as having in some respects anticipated Mendel's point of view.

Maupertuis was even earlier than Kölreuter, his work having been published between 1744 and 1756. He reported on a human pedigree showing polydactylism, and discussed albinism in man and a color pattern in dogs. He also developed a theory of heredity somewhat like Darwin's pangenesis. Glass (1947) has reviewed this work in detail; he sees Maupertuis, in some respects, as a forerunner of Mendel. This is, to my mind, based largely on the interpretation of rather obscure passages in terms of what we now know. In any case, it is clear that Maupertuis had little or no effect on later developments in the study of heredity.

Naudin, a contemporary of Mendel, published his accounts between 1855 and 1869. He studied a series of crosses involving several genera of plants. In several respects he made real advances. Like several of his predecessors, he emphasized the identity of reciprocal hybrids. He also emphasized the relative uniformity of F_1 as contrasted to the great variability of F_2; and he saw the recombination of parental differences in F_2. But there was no analytical approach, no ratios were recognized, and no simple and testable interpretations were presented. The expression "laws of Naudin-Mendel," sometimes seen in the literature, is wholly unjustified.

Mendel's analysis could not have been made without some knowledge of the facts of fertilization—specifically, that one egg and one sperm unite to form the zygote. This was not known until a few years before his time and was not generally recognized even then. Darwin, for example, thought that more than one sperm was needed for each egg, both in animals and in plants.

Direct observations on fertilization had to wait for the development of microscopes. Leeuwenhoek saw animal spermatozoa under a microscope in 1677 and thought that one was sufficient to fertilize an egg— but this was neither directly observed, nor generally accepted, for animals, until two hundred years later (see Chapter 3).* In the lower plants, fertilization was observed by Thuret in 1853 (Fucus), Pringsheim in 1856 (Oedogonium), and De Bary in 1861 (fungi). In seed plants, the work of Amici was especially important. In 1823 he recorded the production of the pollen tube, which in 1830 he traced to the ovary and even to the micropyle. In 1846 he showed that in orchids, there is a cell already present in the ovule, which, inactive until the pollen tube arrives, then develops into the embryo. This work was confirmed and extended by Hofmeister and others, so that Mendel could write in his paper: "In the opinion of renowned physiologists, for the purpose of propagation one pollen cell and one egg cell unite in Phanerogams into a single cell, which is capable by assimilation and formation of new cells, of becoming an independent organism." Nevertheless, there was not general agreement on the point. Naudin (1863) repeated the experiments of Kölreuter and of Gärtner, placing counted numbers of pollen grains on stigmas and concluding that a fully viable seed required more than one grain. It appears, from his let-

---

* Leeuwenhoek also saw conjugation in ciliated Protozoa (1695), but this observation was not understood until the unicellular nature of these animals was made out two centuries later.

ters to Nägeli, that Mendel himself also repeated this experiment (using Mirabilis, as had Naudin) and found that a single grain was sufficient. He did not publish this result, and does not refer to this approach to the question in his paper.

CHAPTER 2

MENDEL

Gregor Johann Mendel was born in 1822 in the village of Heinzendorf in northern Moravia—then a part of Austria, now in Czechoslovakia, near the Polish border. The area had long been populated by people of German and Czech ancestry, living side by side and presumably intermarrying. Mendel's native tongue was the peculiar Silesian dialect of German; in later life he had to learn to speak Czech. He came of peasant stock, and only by persistence and hard work was he able to get a start in education. In 1843 he was admitted as a novice at the Augustinian monastery at Brünn; four years later he became a priest. He took an examination for a teaching certificate in natural science and failed (1850). It has been suggested that the examining board was biased because he was a priest or because his scientific views were unorthodox; the plain fact seems to be that he was inadequately prepared. In order to remedy this, he spent four terms, between 1851 and 1853, at the University of Vienna, where he studied physics, chemistry, mathematics, zoology, entomology, botany, and paleontology. In the first term he took work in experimental physics under the famous Doppler and was for a time, an "assistant demonstrator" in physics. He also had courses with Ettinghausen, a mathematician and physicist, and with Redtenbacher, an organic chemist—both productive research men. We may surmise that this background led to his use of quantitative and experimental methods in biological work. Another of his professors at Vienna, Unger in botany, was also an outstanding figure. Unger was one of the important men in the development of the cell theory; he had demonstrated the antherozoids of mosses and correctly interpreted them as the male generative cells, and he had shown (in opposition to Schleiden) that the meristematic cells of higher plants arise by division. In 1855 Unger published a book on the anatomy and physiology of plants that is rated by Sachs as the best of its time; in this book he made the first suggestion that the fluid content of animal cells and that of plant cells are essentially similar. Mendel was thus in contact with at least two first-rate research scientists, and evidence of their influences upon him shows in his major paper.

Mendel returned to Brünn after the summer term of 1853 at Vienna. At a meeting of the Vienna Zoological-Botanical Society in April, 1854, his teacher Kollar read a letter from him, in which he discussed the pea weevil (*Bruchus pisi*). In the summer of 1854, Mendel grew thirty-four strains of peas; he tested them for constancy in 1855. In 1856 he began the series of experiments that led to his paper, which was read to the Brünn Society for Natural History in 1865 and was published in their proceedings in 1866. Before discussing this paper and its consequences, it will be well to describe some later events in Mendel's life.

He was interested in honeybees and was an active member of the local beekeepers' society. He attempted to cross strains of bees, apparently without success. It has been suggested by Whiting and by Zirkle that he probably knew of the work of Dzierzon on bees, and that Dzierzon's description of segregation in the drone offspring of the hybrid queen may have given Mendel the clue that led to his studies of peas. He is also known to have kept mice, and Iltis and others have suggested that he may have first worked out his results with them, but hesitated, as a priest, to publish on mammalian genetics. These suggestions both seem unlikely to me; there seems no reason to doubt Mendel's own statement: "Experience of artificial fertilization, such as is effected with ornamental plants in order to obtain new variations in color, has led to the experiments which will here be discussed." Perhaps the selection of peas as his experimental material was due in part to Gärtners's account of the work of Knight on peas.

Mendel was also interested in meteorology. At least as early as 1859, he was the Brünn correspondent for Austrian regional reports, and he continued to make daily records of rainfall, temperature, humidity, and barometric pressure to the end of his life. He also kept records of sunspots and of the level of ground water as measured by the height of the water in the monastery well. In 1870 a tornado passed over the monastery, and Mendel published a detailed account of it in the *Proceedings of the Brünn Society*. He noted that the spiral motion was clockwise, whereas the usual direction is counterclockwise. He gave many details, and attempted a physical interpretation. This paper was stillborn, as was his earlier one on peas, published in the same journal. According to Iltis, a catalogue issued in 1917 lists 258 tornadoes observed in Europe but does not include Mendel's account.

In 1868 Mendel was elected abbot of the Brünn monastery. This led to administrative duties and, beginning in 1875, to a controversy with the government on taxation of monastery property. It appears that he continued his meteorological and horticultural observations, but his productive scientific work was finished about 1871. He died January 6, 1884.

Mendel sent a copy of his major paper to Nägeli, together with a letter in which he stated that he was continuing his experiments, using Hieracium, Cirsium, and Geum. Nägeli was professor of botany at Munich and a major figure of his time in biology. He was also interested in heredity and was actively working on it. He completely failed to appreciate Mendel's work and made some rather pointless criticisms of it in his reply to Mendel's letter. He did not refer to it in his publications. He was greatly interested in Hieracium, however, which fact led to a correspondence with Mendel. Nägeli's letters have been lost, but he kept some of Mendel's letters to him. Found among his papers, these were published by Correns in 1905 (I have used the translation in *The Birth of Genetics,* issued in 1950 as a supplement to Volume 35 of *Genetics).* There are ten of these letters, written between 1866 and 1873, and they give a picture of Mendel's biological work during the period. Because of Nägeli's interest, much of the account has to do with Hieracium, the subject of Mendel's only other published paper in genetics (published in 1870 in the *Proceedings of the Brünn Society* for 1869; a translation may be found in Bateson's *Mendel's Principles,* 1909).

The work on Hieracium must have been a great disappointment to Mendel. He obtained several hybrids by dint of much hard work, and all of them bred true. It is now known that this occurs because the seeds are usually produced by apomixis, that is, they are purely maternal in origin and arise without the intervention of meiosis or fertilization (Raunkiär 1903, Ostenfeld 1904). In other words, this was the worst possible choice of material for the study of segregation and recombination—for reasons that could not be guessed at the time.

It appears from Mendel's letters to Nägeli that he was very actively engaged in genetic studies on several other kinds of plants through 1870. His experiments (previously mentioned) with single pollen grains of Mirabilis were repeated in two different years with the same result. He reports studies on Mirabilis, maize, and stocks. Of these three he says "Their hybrids behave exactly like those of Pisum." The character studied in stocks was hairiness; with respect to flower color in this plant, he says the experiments had lasted six years and were being continued—this in 1870. He had grown 1500 specimens for the purpose in that year; his difficulty arose from the mutiplicity of shades that were hard to separate.

In Mirabilis he had seen and understood the intermediate color of a heterozygote and had made the appropriate tests to establish this interpretation. He also mentioned experiments with several other plants— Aquilegia, Linaria, Ipomoea, Cheiranthus, Tropaeolum, and Lychnis.

The picture that emerges is of a man very actively and effectively experimenting, aware of the importance of his discovery, and testing and extending it on a wide variety of forms. None of these results were published; it is difficult to suppose that his work would have been so completely ignored if he had presented this confirmatory evidence, even though it was not enough to convince Nägeli.

This, in outline, is the man. I have tried to give an account of him in order to form a basis for judging his paper—how it came about that he did the work, and what one is to think in view of the analysis by Fisher that will be discussed. A fuller account of Mendel will be found in the biography by Iltis.

There are a number of new procedures in Mendel's work. He himself said in the paper, ". . . among all the numerous experiments made [by his predecessors], not one has been carried out to such an extent and in such a way as to make it possible to determine the number of different forms under which the offspring of hybrids appear, or to arrange these forms with certainty according to their separate generations, or definitely to ascertain their statistical relations." One may agree with Bateson's comment on this passage: "It is to the clear conception of these three primary necessities that the whole success of Mendel's work is due. So far as I know this conception was absolutely new in his day."

This was his experimental approach, but it was effective because he also developed a simple interpretation of the ratios that he obtained and then carried out direct and convincing experiments to test this hypothesis. The paper must be read to be appreciated. As has often been observed, it is difficult to see how the experiments could have been carried out more efficiently than they were.

As Fisher (1936) puts it, it is as though Mendel knew the answer before he started, and was producing a demonstration. Fisher has attempted to reconstruct the experiments as carried out year by year, knowing the garden space available and the number of years involved.* He concludes

---

* Fisher's dates are wrong. He gives them as 1857 to 1864, but it is clear from Mendel's letters to Nägeli that the final year was 1863. Fisher includes the two years of preliminary testing in the eight years that Mendel says the experiments lasted. I have interpreted the statement to mean that these two years preceded the eight years of actual experiments—an interpretation also given by Yule (1902). Fisher's interpretation may be right, but if Yule and I are right there are two more years available and Fisher's year-by-year reconstruction needs revision. It may also be pointed out that Mendel

that the crosses were carried out in the order in which they are described. He also points out several other aspects of the work that seem significant. For example, in testing F_2 individuals to distinguish homozygous dominants from heterozygotes, Mendel must have had a much larger number of seeds illustrating the 3 : 1 ratio than those recorded in F_2; but he did not report these numbers (if he even troubled to count them). Evidently he felt that larger numbers were of no importance.

The most serious matter discussed by Fisher is that Mendel's ratios are consistently closer to expectation than sampling theory would lead one to expect. For yellow vs. green seeds, his F_2 numbers were 6022 : 2001—a deviation of 5 (from 3 : 1), whereas a deviation of 26 or more would be expected in half of a large number of trials, each including 8023 seeds. Fisher shows that this same extremely close fit runs through all Mendel's data. He calculates that, taking the whole series, the chance of getting as close a fit to expectation is only .00007, that is, in only 1 trial of 14,000 would one expect so close an agreement with expectation.

If this were all, one might not be too disturbed, for it is possible to question the logic of the argument that a fit is too close to expectation. If I report that I tossed 1000 coins and got exactly 500 heads and 500 tails, a statistician will raise his eyebrows, though this is the most probable exactly specified result. If I report 480 heads and 520 tails, the statistician will say that is about what one would expect—though this result is less probable than the 500 : 500 one. He will arrive at this by adding the probabilities for all results between 480 : 520 and 520 : 480, whereas for the exact agreement he will consider only the probability of 500 : 500 itself. If now I report that I tossed 1000 coins ten times, and got 500 : 500 every time, our statistician will surely conclude that I am lying, though this is the most probable result thus exactly specified. The argument comes perilously close to saying that no such experiment can be carried out, since every single exactly specified result has a vanishingly small probability of occurring.

In the present case, however, it appears that in one series of experi-

---

probably used some time and garden space in the later years of this period to carry out the experiments with beans and hawkweeds and with the several other plants referred to in the letters to Nägeli.

Fisher also quotes extensively from a paper by Nägeli (1865), and concludes that "it is difficult to suppose that these remarks were not intended to discourage Mendel personally, without drawing attention to his researches." But this paper of Nägeli's was published before Mendel's—clearly before Nägeli could have known anything about Mendel's work!

ments Mendel got an equally close fit to a *wrong* expectation. He tested his F_2 plants that showed dominant characters to see which were homozygous and which were heterozygous, since his scheme required that these occur in the ratio of 1 : 2. For the seed characters (yellow vs. green, round vs. wrinkled), it was necessary only to plant the F_2 seed and observe the seeds the resulting plants produced when allowed to self-pollinate. For the other characters, it was necessary to plant the F_3 seeds and see what kinds of plants they produced. For this purpose, Mendel planted 10 seeds from each tested F_2 dominant. If the tested plant was heterozygous, one-fourth of its offspring would show the recessive. Fisher points out that there is an uncertainty here that was not taken into account by Mendel. For a plant that is heterozygous, the chance that any one offspring will not be a homozygous recessive is .75. The chance that none of 10 will be a homozygous recessive therefore is $(.75)^{10} = .0563$. That is to say, by this test between 5 and 6 percent of the actual heterozygotes will be classified as homozygotes. Fisher shows that Mendel's results are very close to the 2 : 1 ratio expected without this correction and are not in close agreement with the corrected expectation of 1.8874 to 1.1126— in fact as poor an agreement (with the corrected expectation) as Mendel recorded would be expected to occur rather less often than once in 2000 tries.

The argument that a fit to expectation is not close enough is not subject to the criticisms that were levelled earlier against the argument that a fit is too close. There are, however, some further aspects that need discussion. The critical passage in Mendel's paper reads: "Für jeden einzelnen von den nachfolgenden Versuchen wurden 100 Pflanzen ausgewählt, welche in der ersten [second, by current terminology] Generation das dominierende Merkmal besassen, und um die Bedeutung desselben zu prüfen, von jeder 10 Samen angebaut." Fisher is right if only 10 seeds were planted from each tested F_2 dominant. If the experiment included at least 10 seeds but often more than 10, then the correction to the 2 : 1 expectation will be less, and Fisher's most telling point will be weakened. The statement by Mendel seems unequivocal, but the possibility remains that he may have used more than 10 seeds in some or many tests.

There is a possible slight error in Fisher's expectations. In the pea flower, the anthers are closely apposed to the style, and if a plant is allowed to self-pollinate it may be expected that, as a rule, one anther will break at one point. The pollen grains near the break will then be first on the stigma and will be the ones that function. Under these conditions, it may be that the functioning pollen will not be a random sample but will represent all or most of the grains from one or a few pollen-mother-cells. This does not seem likely to be an important factor, since there are so few seeds per flower; but in the limiting case it could result in the sampling error (from a self-pollinated heterozygote) being limited to the eggs alone. Calculations based on this improbable limiting assumption indicate that Fisher's general conclusions would still hold good; but the point remains that in any such analysis one needs to examine the assumptions very carefully, to make sure there may not be some alternative explanation.

Mendel's experiments have been repeated by many investigators, and the question arises: have they also reported unexpectedly close agreement with expectation? For the F_2 ratio for yellow vs. green seeds, the data from several sources have been tabulated by Johannsen, and the statistical calculations have been carried out by him, with the results shown in Table 1.

TABLE 1. F_2 Results, Pea Crosses

Source	Yellow	Green	Total	Dev. from 3 in 4	Prob. Error	Dev. ÷ P.E.
Mendel, 1866	6,022	2,001	8,023	+ .0024	± .0130	.18
Correns, 1900	1,394	453	1,847	+ .0189	± .0272	.70
Tschermak, 1900	3,580	1,190	4,770	+ .0021	± .0169	.12
Hurst, 1904	1,310	445	1,775	− .0142	± .0279	.51
Bateson, 1905	11,902	3,903	15,806	+ .0123	± .0093	1.32
Lock, 1905	1,438	514	1,952	− .0533	± .0264	2.04
Darbishire, 1909	109,060	36,186	145,246	+ .0035	± .0030	1.16
Winge, 1924	19,195	6,553	25,748	− .0180	± .0125	1.44
Total	153,902	51,245	205,147	+ .0008	± .0038	.21

Source: Johannsen, 1926.

Evidently this is in good agreement with expectation. It would be expected that the values in the last column would be more than 1.00 in half of the series, less than 1.00 in half—which happens to be just what is observed. One observer, Tschermak, achieved an even closer approach to 3 : 1 than did Mendel. Of the eight observers, five (including Mendel) obtained a small excess of dominants, three got a small deficiency. The poorest fit (that of Lock) would be expected to occur in about 1 out of 6 tries, and it did occur in 1 of 8 series. The over-all impression is that the agreement with expectation is neither too good nor too poor.

In summary, then, Fisher's analysis of Mendel's data must stand essentially as he stated it. There remains the question of how the data came to be as they are. There are at least three possibilities:

1. There may have been an unconscious tendency to classify somewhat doubtful individuals in such a way as to fit the expectation.

2. There may have been some families that seemed aberrant, and that were omitted as being probably due to experimental error.

3. Some of the counts may have been made for him by students or assistants who were aware of his expectations, and wanted to please him.

None of these alternatives is wholly satisfactory, since they seem out of character, as judged by the whole tone of the paper.

Perhaps the best answer—with which I think Fisher would have agreed—is that, after all, Mendel was right!

CHAPTER 3

1866 TO 1900

The period between the publication of Mendel's paper and its rescue in 1900 from oblivion was dominated by the development of the theory of evolution and its implications. So far as heredity was concerned, it was largely a period of the production of theories. There were, however, several real advances which helped to make Mendel's results acceptable. Here we may mention the germ-plasm theory with its emphasis on the effects of germinal material on the body rather than the reverse; the resulting challenge of the inheritance of acquired characters; the striking increase of knowledge of the cytological details of fertilization and cell division; and the increasing emphasis on the importance of discontinuous variation. This chapter will be concerned chiefly with these topics.

The outstanding figure of the time was August Weismann (1834–1914), who was professor of zoology at Freiburg for many years. From 1862 to 1864 he published several papers on the embryology of Diptera, and these seem to have led to much of his later theoretical work.

In these flies, the so-called pole cells are set aside in early cleavage divisions, subsequently to develop into the germ cells. Their early separation from the somatic cells, and their relatively independent development, seem to have suggested the germ-plasm theory to Weismann (1883), although he was also aware of a similar idea expressed by Nussbaum. According to this scheme, the germ line is the continuous element, and the successive bodies of higher animals and plants are side branches budded off from it, generation after generation. This is, of course, only a way of looking at familiar facts, since Weismann recognized that in the higher plants and in many animals the visible distinctness of the germ line only appears late in development and, in fact, that many cells that will not normally give rise to germ cells still retain the potentiality of doing so. Nevertheless, the idea was a fruitful one, since it led to an emphasis on the effects of the hereditary material on the soma and to a minimizing of effects in the reverse direction—a point of view already foreshadowed by Aristotle. This in turn led to a challenging of the hypothesis of the inheritance of acquired characters, which had already been questioned by Du Bois Reymond in 1881. But it was the writings of Weismann that really showed that the hypothesis was unnecessary and improbable, and that the supposed evidence for it was weak. In the special case of the inheritance of mutilations, already questioned by Aristotle and Kant, Weismann carried out an experiment. He cut off the tails of mice for twenty-two successive generations and found no decrease in the tail length at the end of the experiment.

Weismann suffered from eye trouble, and finally had to give up microscopic work and experimental studies,* although he kept the latter going in his laboratory through the work of students and assistants. His own work, however, came to be largely theoretical. He was in close touch with the activity of the time in the cytological study of the chromosomes and played a large part in the theoretical developments in that field—which may be discussed conveniently at this point.

The importance of the nucleus in the cell theory had gradually become evident, though not universally recognized; but with the observations of O. Hertwig (1875) and of Fol (1879) on the fertilization of the egg of the sea urchin, the role of the nucleus in fertilization and cell division was placed beyond doubt.

There followed a few years (about 1882 to 1885) during which a whole series of investigators laid the foundations of our knowledge of the behavior of the chromosomes in mitosis and meiosis. This rapid development seems to have been due mainly to two events. The microtome was improved at about this time (by Caldwell), and made possible the production of serial sections of uniform thickness, thereby improving the quality of microscope preparations. At about the same time, van Beneden discovered the advantages of Ascaris for the study of chromosomes, and this animal (the threadworm of the horse) became one of the standard objects for such studies.

During these few years Flemming and Strasburger recognized chromosomes. Van Beneden showed that the daughter halves of the mitotic chromosomes pass to opposite poles at mitosis; that, in Ascaris, the fertilized egg receives an equal number of chromosomes from each parent; and that the meiotic divisions result in halving the number present in the fertilized egg. Here, then, was the first demonstration of the double na-

---

* Both Mendel and Correns also suffered from eye trouble brought on by excessive work with strong light.

ture of the soma and the simplex condition of the germ cells (a relation deduced by Mendel from his genetic results but not recognized by his contemporaries). It had, however, been anticipated by Weismann, who supposed that the function of the polar body divisions in the egg was to prevent an indefinite accumulation of ancestral hereditary units and predicted that a similar reduction would be found in the formation of the sperm.

In 1883 there appeared a remarkable essay by Roux, in which he argued that the linear structure of the chromosomes and their point-by-point division into equal longitudinal halves were such striking and widespread phenomena that they must have some selective value. This, he suggested, lay in their effectiveness in assuring that each daughter cell received the same complement of chromosomal material. He saw this as a strong argument in favor of identifying the chromosomes as the bearers of the units of heredity. These units were also here first specified as being arranged in a linear series—the visible slender strands of the dividing chromosomes.

Roux applied these ideas to the cleavage divisions of the fertilized egg of the frog. He was the "father" of experimental embryology and had carried out experiments which he believed had shown that the two cells arising from the first division are equivalent, but that the second division leads to differences in the potentialities of the daughter cells.* He therefore concluded, in the 1883 essay, that at the second division the process of mitosis does not lead to exactly equal complements of hereditary units in the daughter cells. This was the beginning of the hypothesis that differentiation is due to somatic segregation—the sorting out of hereditary elements at somatic cell divisions.

These ideas were at once adopted by Weismann, who elaborated them into an intricate theory of heredity and development. According to this scheme, the chromosomes are the bearers of the hereditary material. Weismann supposed that each chromosome remains intact in successive generations, and is simply passed on through the germ line from generation to generation. Since an individual may resemble several different ancestors in one respect or another, he concluded that each chromosome carries all the hereditary elements necessary to produce a whole individual. The different chromosomes of an individual may have been derived from many different ancestral lines, and they therefore differ among themselves. Each is potentially able to determine the characteristics of a whole organism, but in the development of a particular part, only one

---

* Later experiments have not borne out this conclusion about the second division.

chromosome is effective at any given time and place. There is, in a sense, a competition between the various chromosomes, and the nature of each characteristic depends on the outcome of this competition at each critical time and place in the developing embryo. Each chromosome was supposed to be made up of smaller units, and these in turn of still smaller subunits. These were distributed unequally at somatic divisions, forming the basis of differentiation.

This theory was elaborated in great detail and was widely known and discussed, but it was not accepted in detail, because it was so hypothetical and seemed to offer so little basis for experimental testing. There can be no question of the importance and widespread influence of much of Weismann's work, but the elaboration of his scheme of heredity and development led to widespread resistance to even the sounder parts of his interpretations. As Wilson expressed it in 1900:

> Weismann's . . . theories . . . have given rise to the most eagerly contested controversies of the post-Darwinian period, and, whether they are to stand or fall, have played a most important part in the progress of science. For, aside from the truth or error of his special theories, it has been Weismann's great service to place the keystone between the work of the evolutionists and that of the cytologists, and thus to bring the cell-theory and the evolution theory into organic connection.

One of the workers of the time who was greatly influenced by Weismann but was unwilling to accept all of his conclusions, was de Vries. In his *Intracellular Pangenesis* (1889), de Vries developed a theory of heredity different from Weismann's. He pointed out that there are two parts to Darwin's hypothesis of pangenesis—the view that there are persistent hereditary units which are passed on through successive generations, and the view that these are replenished by gemmules derived from the somatic tissues. Following Weismann and others, de Vries rejected the second of these views, but he retained the first. This might have led to an interpretation like Weismann's, but de Vries added an essential point, namely, that the units (which he called "pangens") are each concerned with a single character, and that these units may be recombined in various ways in the offspring. This was a clear approach to the Mendelian point of view, and helps to explain why, eleven years later, de Vries was one of the three men who discovered and appreciated Mendel's paper.

There was a difficulty about Darwin's views on the effectiveness of natural selection, if one supposed that most characters blend in hybrids, and that it is just these characters that are important in selection, either natural or artificial. The difficulty is that a favorable variation will, on this basis, be rapidly diluted by crossing to the parental form, and systematic change of the whole population will be painfully slow, if possible at all. It was not until well after 1900 that this difficulty (the "swamping effect") was cleared up, as we shall see in Chapter 9. But it led to an increasing interest in "sports," which, as Darwin had realized, showed little tendency to "blending" or "swamping." This interest is apparent in the writings of Galton as early as 1875.

Bateson, in *Materials for the Study of Variation* (1894), expressed the growing dissatisfaction with the view that selection was a sufficient explanation of evolution. He felt that too little was known about the facts of variation, and that the current phylogenetic theories were of little value. As he put it:

> In these discussions we are continually stopped by such phrases as "if such and such a variation then took place and was favorable" or "we may easily suppose circumstances in which such and such a variation if it occurred might be beneficial," and the like. The whole argument is based on such assumptions as these—assumptions which, were they found in the arguments of Paley or of Butler, we could not too scornfully ridicule. "If," we say with much circumlocution, "the course of Nature followed the lines we have suggested, then, in short, it did." That is the sum of our argument.

This dissatisfaction with the then-current views led Bateson, Korschinsky, and de Vries to lay great emphasis on the importance of discontinuous variations. As we can now see, they overemphasized the distinction between the two kinds of variation; but the immediate result was to focus attention on sharply separable variations, and these were more easily susceptible of exact study. Again, it was no accident that de Vries was one of the discoverers of Mendel's paper, and that Bateson was perhaps the most important of the early advocates of the Mendelian approach.

During the period in question, a quite different approach to the study of heredity was developed by Francis Galton. Galton, who was a cousin of Darwin, had carried out an experiment to test the theory of pangenesis. He performed extensive blood transfusions between different strains of rabbits and found no effects on their descendants in either the first or the second generation. Darwin admitted that he would have expected effects but felt that his gemmules were not necessarily to be expected in the blood, since the theory was supposed to apply even to organisms without a circulatory system. Galton agreed that the experiment was not decisive.

Galton's theoretical contribution arose from his feeling for the importance of quantitative study. He felt that almost anything could be measured. He attempted to develop a quantitative scale for beauty; and he carried out a study on the effectiveness of prayer, by examining mortality rates for crowned heads (whose subjects prayed for their health), and by comparing the frequencies of shipwreck for vessels that did and did not carry missionaries. Like Mendel, he also studied meteorology.

He developed the idea of correlation as a result of tabulating the relationship between the height of parents and that of offspring in human families. He saw what was needed in geometrical terms and referred the algebraic problem to the mathematician Dickson, who then produced the regression coefficient.* Galton used this to give a simple numerical value for the degree of resemblance between parents and offspring, thus initiating a whole field of study. He tabulated a large series of data on the colors of pedigreed basset hounds, and based his *Law of Ancestral Inheritance* on the results. These results showed that, on the average, an individual inherits ¼ of his characteristics from each parent, ¹⁄₁₆ from each grandparent, ¹⁄₆₄ from each great-grandparent, and so on. This ingenious approach was followed by many of his successors, but failed to give the hoped-for insight into the mechanisms involved. As it happened (through no fault of Galton's), it led to a long and bitter controversy that wasted much time and printer's ink in the early years of this century (Chapter 9).

The question has often been raised: Would any biologist have appreciated Mendel's work if he had seen the paper before 1900? My own candidate for the most likely person to have understood it is Galton, because of his interest in discontinuous variation, his mathematical turn of mind, and his acceptance of Weismann's view that the hereditary potentialities of an individual must be halved in each germ cell.

One of the "eager controversialists" referred to by Wilson was Haacke, who published a series of anti-Weismann papers between 1893 and 1897. These papers, which have been overlooked by many of the more recent authorities (but not by all—see Correns, 1922), contain the nearest approach to the Mendelian interpretation before the rediscovery. They make difficult reading, because the results and conclusions are so buried in a mass of polemics.

Haacke crossed normal albino mice with waltzing mice that were colored. In F_1 he got only colored normals, and in F_2 (which he called the "third generation," the parent strains being considered the first) he rec-

---

* This account is from Galton's autobiography. It appears that the correlation coefficient had already been developed by Bravais in 1846.

ognized the occurrence of the recombination types. The analysis is based on his supposition that all structural characters (including the waltzing habit, presumably due to some structural change) were inherited through the centrosomes, and all chemical characters (that is, coat color in this case) through the nucleus. There follows a surprisingly modern-sounding hypothetical scheme. He designates the plasma (= centrosomes) of the waltzer as t (for *Tanzmaus*) and that of the nonwaltzer as k (for *Klettermaus*); and the nucleus of the albino as w, that of the colored mouse as s. He specifies that, at the reduction division, t separates from k, and w from s, resulting in four kinds of eggs or of sperm: ts, tw, ks, and kw. Fertilization of these will then result in sixteen kinds of individuals, which he lists: ts, ts; ts, tw; ts, ks; and so on to kw, kw. He points out which of these will breed true and which will not—in other words, a straightforward Mendelian analysis for two pairs of genes—except that no ratios are given. Here he makes what at first glance is an astonishing statement: "Ob die Anzahl der Chromosomen bei den Mäusen bekannt ist, weiss ich nicht, man würde daraus die möglichen Kombinationen aufstellen können." This statement sounds at least ten years ahead of the thinking of the time—but study of the context indicates that Haacke was misled by the Weismannian idea that each chromosome contains all the hereditary material needed to produce an individual, and he needed the chromosome number to calculate the probability of getting all chromosomes in a gamete purely maternal or purely paternal. He was unaware of the 1 : 1 segregation in heterozygotes and, in fact, apparently visualized various kinds of heterozygotes, at least for color.

Only after this hypothetical analysis are we told that he had raised over 3000 mice in his experiments, and that they were in "the most beautiful agreement" with the theories he developed. No numbers are given, and no ratios.* He does, however, insist that the separation of t from k and of w from s must be complete, since extracted waltzers and albinos both breed true when mated to their own kind.

The paper from which this summary has been abstracted appeared in one of the best-known biological journals of the time (*Biologisches Centralblatt*, vol. 13, 1893), so that it is difficult to see why it was overlooked so long. One reason is surely the polemic nature of the paper, which led to the data and conclusions being emphasized primarily as ammunition against Weismann rather than for their own sake.† Another

---

* The actual data were published much later, well after the rediscovery of Mendel's work *(Arch. Entw.-mech.,* 1906).

† One gibe at Weismann is perhaps worth citing. Haacke argued that Weismann might have gotten further if he had made crosses between the various kinds of fancier's

reason is their use to support the unpopular (and erroneous) idea of the genetic importance of the centrosomes. Finally, the failure to give actual counts made the data seem as speculative as the discussion in which they are imbedded.

mice to study the inheritance of coat colors, "instead of cutting off the tails of his unfortunate mice and those of their children and of their children's children unto the twentieth generation."

CHAPTER 4

THE REDISCOVERY

To Bateson and to de Vries, the logical approach to the study of heredity seemed to be the study of variation, which was then to be followed by the study of the transmission of variations. As the event showed, the effective approach was the reverse of this, since the origin of variability could begin to be analyzed only after the nature of segregation and recombination was understood.

By the end of the century both men felt that the time had come to begin a serious study of the inheritance of discontinuous variations. In 1899 Bateson published an analysis of what was needed, which is remarkable, among other things, for the statement "If the parents differ in several characters, the offspring must be examined statistically, and marshalled, as it is called, in respect to each of those characters separately." Here was, clearly, a man whose mind was ready to appreciate the Mendelian approach.

The story of the finding of Mendel's paper and of the confirming of his results in 1900 has often been told—perhaps most fully by Roberts (1929).

Mendel had forty reprints of his paper. He sent copies to Nägeli and to Kerner, professors of botany at Munich and at Innsbrück, respectively, and both interested in plant hybrids. It is not known what happened to the other thirty-eight copies; after Kerner's death, his copy was found in his library with the pages uncut.* As was pointed out earlier, Nägeli did reply but did not appreciate the work or refer to it in print. The journal was perhaps rather obscure, but the Brünn Society had a considerable exchange list, and its *Proceedings* were sent to more than 120 libraries. According to Bateson, there were at least two copies in London. Only

---

* As will appear below, a third reprint was in the library of the Dutch botanist Beijerinck. I have received, through the kindness of Dr. H. Gloor and Dr. F. Bianchi, a photostat of the cover and first page of this reprint; there is no indication of how or when Beijerinck acquired it, or to whom Mendel sent it.

four printed references to the paper before 1900 are known, however, other than a listing in the *Royal Society Catalogue* of scientific papers. Hoffmann (1869) published an account of experiments with beans, in which Mendel's paper is referred to without any indication of its nature. Focke (1881) published a rather extensive account of the literature of plant hybridization, in which he referred to Mendel's paper under the heading "Pisum." He failed to appreciate or even to understand the work, but he did state that Mendel "believed that he found constant numerical relationships between the types"—a statement that ultimately led to the paper being found, as will be discussed.

The third reference was by L. H. Bailey (1895), who copied Focke's statement without having himself seen Mendel's paper; this was the source that led de Vries to Mendel, according to one account. Finally, Mendel was listed, without comment, as a plant hybridizer by Romanes in the ninth edition of the *Encyclopaedia Britannica* (1881–1895)—evidently again following Focke.

Hugo de Vries (1848–1935) was born in Holland. His university training was largely in Germany, where he studied plant physiology with Sachs. In 1871 he became a lecturer at the University of Amsterdam and, from 1881 until his retirement, was a professor there. His early work was on local floras, the microorganisms in water supplies, and the turgor of plant cells. In the latter field, he carried out a beautiful series of quantitative studies of the effects of the concentrations of various salts on plasmolysis. These results were of importance in the development (by Arrhenius and van't Hoff) of the ionic theory of the osmotic properties of solutions of electrolytes. His *Intracellular Pangenesis* (1889) has been described in Chapter 3; his work on mutation will be discussed in Chapters 10 and 11.

De Vries published three papers on Mendelism in 1900, one of which has, for the most part, been overlooked. The first was read by G. Bonnier before the Paris Academy of Sciences on March 26 and was published in the Academy's *Comptes Rendus.* A reprint of this paper was received by Correns on April 21. Another paper by de Vries is dated "Amsterdam, March 19, 1900," and was published in the *Revue général de botanique,* which was edited by Bonnier. It seems likely, then, that these two French manuscripts were sent to Bonnier at the same time. The third paper, in German, was received by the editor of the journal *(Berichte der deutschen botanischen Gesellschaft)* in Berlin on March 14 and was published April 25. These dates are of some interest because the brief note in the *Comptes Rendus*, the first to be published, does not mention Mendel, though it uses some of his terminology. The *Revue général* paper is the one that is rarely cited. It is longer and does mention Mendel—though only on the last page, where is also an added footnote referring to the *Berichte* paper and to the papers by Correns and by Tschermak, which did not appear until May (apparently this paper was published in July). The reference to Mendel on this page may be translated as follows: "This law is not new. It was stated more than thirty years ago, for a particular case (the garden pea). Gregor Mendel formulated it in a memoir entitled *'Versuche über Pflanzenhybriden'* in the *Proceedings of the Brünn Society.* Mendel has there shown the results not only for monohybrids but also for dihybrids.

"This memoir, very beautiful for its time, has been misunderstood and then forgotten."

In the *Berichte* paper, the second to be published but evidently the first to be submitted, there is much the same material as in the longer French one, but Mendel is mentioned in several places in the text and is given full credit for his discovery.

In a letter received and published by Roberts, de Vries later stated that he had worked out the Mendelian scheme for himself, and was then led to Mendel's paper by reading Bailey's copy of Focke's reference. In 1954, nineteen years after the death of de Vries, his student and successor Stomps reported that de Vries had told him that he learned of Mendel's work through receiving a reprint of the 1866 paper from Beijerinck, with a letter saying that he might be interested in it. This reprint is still in the Amsterdam laboratory, as has been stated.

There is a persistent and widespread story to the effect that de Vries at first intended to suppress any reference to Mendel and changed his mind only when he found that Correns (or Tschermak) was going to refer to him. This is based on the failure to refer to Mendel in the *Comptes Rendus* paper, the first to be published—and, one may add, also in the *Revue général* paper until the last page that was, at least in part, added some months later. This view can be maintained only if it is supposed that the *Berichte* paper was extensively altered in proof—a suggestion that gets some support from the fact that nine of the twenty-two *errata* listed at the end of the volume concern just the pages that would have had to be altered. These *errata* are rather minor, but they do make one wonder if the printer was confused by extensive alterations in the proofs.

A careful comparison of the available dates, however, makes it seem impossible that such changes could have been a result of a letter from Correns after he had seen the *Comptes Rendus* paper, and very unlikely also that a letter from Tschermak could have been involved. Both of these men have stated (Roberts) that they learned that de Vries had the interpretation when they received reprints of this paper from him.

It remains possible, though, that de Vries did come to realize that Correns knew and appreciated Mendel's paper, from the reference in the January, 1900 paper on xenia (to be discussed). This conclusion cannot be accepted as established but seems to be the simplest interpretation of the puzzling facts.

In these three papers de Vries recorded a series of quite different genera of plants that had given the 3 : 1 ratio, and, in several of them, he had also seen the 1 : 1 ratio on crossing the F_1 to the recessive. There was, therefore, no question that the scheme was generally applicable. De Vries concluded that it probably held for all discontinuous variations.

Carl Correns (1864–1933) was a student of Nägeli and of the plant physiologist Pfeffer, who, like de Vries, was a student of Sachs. Correns studied the anatomy and the life cycle of mosses and also became interested in the origin of the endosperm. This tissue in the seeds of higher plants was long supposed to be of purely maternal origin, but it was often observed—especially in maize—that the nature of the endosperm was influenced by the pollen. Correns set out to study this phenomenon (called xenia by Focke). He reached the conclusion that the endosperm was in fact derived from the "double fertilization" that had just been described by Nawaschin in lilies. A preliminary account of these results appeared *(Berichte deutsch. botan. Gesellsch.* January 25, 1900), the manuscript having been received December 22, 1899. The same conclusion was also reached by de Vries in 1899.

In the last paragraph of this paper, Correns pointed out that the superficially similar phenomenon in the case of green and yellow peas was due to color in the cotyledons, that is, in true embryonic tissue "as already correctly pointed out by Darwin and by Mendel." This was the first printed indication that anyone had understood any part of Mendel's work.

In connection with this study, Correns grew hybrids of maize and of peas through several generations, and arrived at the interpretation (that is, the Mendelian one) in 1899. This caused him to read Mendel's paper, because he found Focke's statement that Mendel believed he had found "constant numerical relationships." Correns (in a letter quoted by Roberts) compared his own and Mendel's solution of the problem: ". . . through all that in the meantime had been discovered and thought (I think above all of Weismann), the intellectual labor of finding out the laws anew for oneself was so lightened that it stands far behind the work of Mendel."

Correns reported in detail on his work with peas, in a paper that appeared in May 1900, after he had seen de Vries' account. He fully confirmed Mendel* and said that he had observed the same kind of results with maize; these results were published in full in 1901. He disagreed with de Vries in that he thought there were cases that did not conform to the Mendelian scheme. The only one described in any detail, having to do with the color of the seed coat in peas, seems to have involved the carrying of a dominant gene for color pattern in a plant which also had a recesssive that prevented all color, with the result that the F_1 did not resemble either parent. This type of situation later became familiar but seemed then to contradict the Mendelian scheme.

Erich von Tschermak (1871–1962) was a grandson of Fenzl, under whom Mendel studied systematic botany and microscopy at Vienna. Tschermak was trained at Halle, where he received his doctorate in 1895. His interests were in practical plant breeding, and this led to studies (at Ghent and Vienna) on the effects of crossing and inbreeding on vegetative vigor, following the work of Darwin. In this connection he made crosses of peas and raised F_2, noting the 3 : 1 ratios and also the 1 : 1 on back-crossing to the recessive parent. He later wrote (to Roberts) that he had realized the significance of this result before he found Mendel's paper (through the reference by Focke), but since he had reared only two generations when he published his accounts, he cannot have known that the recessives bred true or that there were two classes of individuals in F_2 that had the dominant character. He published two papers on the subject in 1900. Of these, the first is much less clearly indicative of a real understanding of the situation than is the second, which was written after he had seen the de Vries and Correns papers in the *Berichte*.

William Bateson (1861–1926) was trained as a zoologist at Cambridge. He was influenced by Sedgwick, by F. Balfour, and by his contemporary, Weldon. The summers of 1883 and 1884 were spent at Hampton, Virginia, and Beaufort, North Carolina, studying the embryology of Balanoglossus under W. K. Brooks. Bateson has recorded that it was Brooks who gave him the idea that heredity is a subject worth studying for itself. In passing, it may be remembered that Brooks also influenced the history of genetics through the fact that both E. B. Wilson and T. H. Morgan were trained by him.

---

* This confirmation included extensive tests over several generations, showing that extracted homozygotes bred true. This was perhaps the one of Mendel's observations that was hardest to accept at the time. We know that Nägeli balked at it, and that as late as 1910, Morgan tried to explain it away.

Bateson's *Materials for the Study of Variation* (1894) and his outline of what was needed in the study of heredity (1899) have been discussed in Chapter 3 and earlier in this chapter. In May, 1900, he read a paper before the Royal Horticultural Society in London, in which he described the work of Mendel and its confirmation by de Vries. According to Mrs. Bateson (1928), he first learned of Mendel's work on a train, while going from Cambridge to London to deliver this paper, and was so impressed by it that he immediately incorporated it into his lecture.

Bateson at once became the most active proponent of the new approach and developed a very active group of workers at Cambridge, including, in the early years of this century, Saunders, Punnett, Durham, Marryatt, and others. Mendelian studies were actively pursued in Germany by Correns, in Austria by Tschermak, in France by Cuénot, and in the United States by Castle and Davenport. These workers and others soon built up a great mass of data and laid the foundations for later developments.

There were two immediate problems: How widespread are the Mendelian phenomena, and what is the interpretation of the so-called compound characters? Another question, that now seems less important, concerned the generality of the phenomenon of dominance.

Mendel's work established his principles for peas and beans; they were confirmed for peas by Correns and by Tschermak. In his 1900 papers, de Vries showed that the principles applied to about a dozen widely different genera of seed plants, including one monocotyledon (maize, which Correns confirmed). There was thus clear evidence for the general applicability to angiosperms. Correns suggested in 1901 that the principles applied to animals, citing a number of experiments from the earlier literature that appeared to be Mendelian. That the principles do apply to animals was definitely shown in 1902, independently by Cuénot (mice) and by Bateson (fowl). It was therefore concluded that the same general scheme must apply to all higher animals and plants; the later applications to invertebrates and to lower plants were, when they were made, interesting chiefly because they offered means of studying new kinds of problems.

The other important question during the early development was that of the nature of "compound characters"—or, as we should now say, cases where more than one pair of genes affect the same character.

Mendel reported a cross with beans, using as parents a strain with white flowers and one with colored flowers. F_1 was colored, but the cross was between quite different species, and these F_1 plants were only slightly fertile. Mendel obtained a total of 31 F_2 plants, of which only one had white flowers. He suggested, tentatively, that there were two independent dominant genes, A_1 and A_2 either of which alone would give colored flowers. The expectation in F_2 would then be 15 colored : 1 white; but because of the small number of plants available he did not urge this interpretation. In 1902, Bateson criticized this suggestion, since he felt that A_1 and A_2 would be expected to behave as alleles; that is to say, he was still thinking in terms of a single gene for each visible character. A similar point of view is found in the 1900 paper by Correns, as has been pointed out, in his argument that the Mendelian principles could not be general; the same view later led Cuénot to lay no emphasis on his demonstration of multiple alleles (see Chapter 8).

Several examples of a related sort soon turned up—first in the case of flower color in the sweet pea, where a cross of two whites gave colored flowers in F_1, and in F_2, 9 colored : 7 white (Bateson and Punnett). Tests of individual F_2 plants showed that this occurred because color development requires the presence of both of two independent dominants. The F_2 ratio is $9 : 3 : 3 : 1$, with the last three classes indistinguishable, rather than the first three, as in Mendel's beans. With this cleared up (by Bateson and Punnett), it was not difficult to interpret the $9 : 3 : 4$ cases that were soon found.

The case of combs in fowl was puzzling at first. Here the familiar "rose" comb was soon shown to be dominant to single, and "pea" was also shown to be dominant to single. When rose was crossed to pea, a new type, called "walnut," appeared; this did not seem to be, structurally, a combination of rose and pea. The relations became clear when it was found that the F_2 ratio is 9 walnut : 3 rose : 3 pea : 1 single (again by Bateson and Punnett)—for Mendel had already shown that this ratio meant that two independent pairs of genes were segregating.

The occasional occurrences of cases in which the heterozygote is intermediate, that is, strict dominance is absent, was indicated by Correns in a footnote added to his 1900 paper in proof. This had been seen and understood, with the appropriate tests carried out, in Mirabilis by Mendel, according to his letters to Nägeli, and is implied in his account of flowering time in peas in the 1866 paper. The Mirabilis case and that of the Andalusian fowl became clear very early in the period after 1900, and as a result, the phenomenon of dominance was recognized as of only secondary importance.

New terminology was needed. Many of the now-familiar terms were introduced by Bateson—such as *genetics,* for the subject itself, and *zygote,* for the individual that develops from the fertilized egg, as well as for the fertilized egg itself (which was the older usage). *Homozygote,* *heterozygote,* and the adjectives derived from them followed. Mendel had spoken of the hybrids and the first generation from them; Bateson suggested that these be designated "F_1" and "F_2," respectively—to stand for first and second filial generations.[*] The term *allelomorph* (later, especially in the United States, shortened to *allele*) also dates from Bateson's early work. Mendel usually used the word *Merkmal* for what we now term *gene,* and this was translated as *character,* often appearing as *unit character*; Bateson usually used the word *factor.* It was somewhat later (1909) that Johannsen introduced the word *gene.*

---

[*] The confusion existing may be illustrated further by the fact that Haacke (Chapter 3) considered the parental generation as the first and referred to what we now call "F_1" and "F_2" as the second and third generations, respectively.

CHAPTER 5

GENES
AND CHROMOSOMES

For many years the standard authority on the chromosomes was Wilson's *The Cell in Development and Inheritance*. The second edition of this work was published in 1900; it gives a full account of the state of knowledge and of current theories about chromosomes at the time of the discovery of Mendel's paper.

The constancy of chromosome number for a species was known, and it was known that this number was usually even, equal numbers coming from the egg and from the sperm. It was known that each chromosome divides longitudinally at each somatic division, and that this division is initiated by an equal division of each visible granule along the length of the chromosome. It was also known that the reduction in chromosome number is accomplished by the last two divisions before the production of the mature gametes (in animals) or gametophytes (in plants). Further, it was generally supposed that the chromosomes are the bearers of the essential hereditary material.

There were, however, a number of things, now part of common biological knowledge, that were not known. It was generally supposed that, when the chromosomes reappear at the end of the resting stage, they first do so as a single continuous thread, or spireme, which then breaks into the number of chromosomes characteristic for the particular species. It had been postulated by Rabl and by Boveri that the chromosomes "do not lose their individuality at the close of division, but persist in the reticulum of the resting nucleus." Although Weismann adopted this view, Wilson felt that it was far from proved. The details of chromosome reduction at meiosis were not at all clear, chiefly because the two-by-two pairing in meiotic prophase was not recognized. The whole idea of definite pairs of chromosomes was missing. It was not recognized that there are different kinds of chromosomes in a single cell. In short, one chromosome was tacitly assumed to be essentially like any other in the species, and in Weismann's writings this assumption was explicitly made.

It was supposed that the bivalent chromosomes in the first meiotic division were condensed from a single continuous spireme and were therefore attached end to end from the beginning; the reduction in number must therefore arise from a transverse division. There were, however, good descriptions indicating that these bivalents divided by means of two successive *longitudinal* divisions in some species; this appeared as a paradox, since it seemed to contradict the view that the reduction is a qualitative one, and not merely quantitative.

These matters were gradually cleared up by the cytologists. That the chromosomes occur in distinct pairs, which can sometimes be recognized by their sizes and shapes, was first indicated by Montgomery in 1901 and was shown conclusively (in a grasshopper) by Sutton in 1902. Both authors showed that one member of each pair was maternal in origin, the other paternal; this interpretation was very soon generally accepted. But both men still thought that the two members of a pair were attached end to end.

The interpretation of the bivalents in the first meiotic division as resulting from side-by-side pairing of separate chromosomes was suggested by Winiwarter in 1901, as a result of his studies of the ovaries of the rabbit. He discussed the difficulties just outlined and concluded that such side-by-side pairing was the simplest way of reconciling the apparent contradictions. This view was not at once generally accepted but slowly gained ground as more and more cytologists saw figures consistent with it. It was not important in the earliest work on the relation between genes and chromosomes but, as will appear, was essential for later developments.

In 1902 Boveri issued a remarkable paper on the results of polyspermy in the fertilization of the eggs of the sea urchin. He showed that, if an excess of sperm is used, two sperm may enter a single egg. Each sperm centriole then divides, and 3-poled or 4-poled spindles result. These eggs may divide into three or four cells at the first cleavage division. The three haploid sets of chromosomes (one from the egg and one from each sperm) divide and pass to these daughter cells more or less at random, so that most cells receive abnormal numbers of chromosomes. These cells usually divide normally for a few divisions, but the resulting embryos are quite abnormal, often appearing to be mosaics, with some portions reasonably normal and others aborted or quite abnormal.

If the four cells resulting from the first two cleavages of a normally fertilized (monospermic) egg are separated, each will give a normal (though small) embryo. Boveri found that if the three or four cells produced by the first cleavage of dispermic eggs were separated, they would sometimes develop into normal embryos—but never would all of the cells from any one egg do so. He showed that this result could not be explained by the number of chromosomes present, since in cells that failed to develop this was often greater than the haploid number—and a single complete haploid set was already known, from other experiments, to be enough for normal development. The results can only be explained on the assumption that the chromosomes differ in their effects on development, and that a cell will not give rise to a normal embryo unless it has at least one complete haploid set of chromosomes.

The paper has a footnote, in which Boveri points out that he, like Weismann, had previously supposed that the chromosomes of an individual were equivalent one to another. This view he now finds untenable, and therefore Weismann's interpretation of chromosome reduction at meiosis must now be revised. This same footnote contains a statement that may be freely translated thus: "I shall consider in another place these and related problems, such as the connection with the results of the botanists on the behavior of hybrids and their offspring." Before this further discussion appeared, the whole matter was clearly analyzed by Sutton, but there can be no doubt that Boveri was near to the solution.

In 1900 Correns had already raised the question of where the Mendelian segregation occurs and had discussed it in several papers, the fullest account being in 1902. He knew, from his experiments with maize hybrids, that the embryo and the endosperm of a given seed are alike. It follows that the three maternal nuclei involved in the double fertilization are alike, as are the two from the pollen. He concluded that segregation is accomplished at latest by the time the megaspore is produced, and that it occurs in the anther at some time before the final division that produces the two sperm nuclei in a single pollen grain. On the other hand, the earliest time it can be supposed to occur is after the sexual organs are formed, as both the ovaries and the anthers of heterozygous plants produce both kinds of gametes. Since the ratio of dominant to recessive gametes is so very close to 1 : 1, he concluded that segregation must come very late in the development—in order to avoid chance (or selective) differences in multiplication of the products. He was, of course, also aware that the last two sporophyte divisions (that is, the meiotic ones) are of a different type and lead to haploid nuclei. He therefore concluded that segregation occurs in these two divisions in the ovule.

In the anther he recognized that the same reasoning would lead to the view that the meiotic divisions were also those that produce segregation.

This should mean that the pollen grains of a heterozygote were of two kinds, in equal numbers. He tested this deduction. He found two flower colors in Epilobium that gave Mendelian results and were associated with differences in the color of the cell sap of the pollen grains. He examined the pollen of the heterozygote and found it to be uniform in color. An exactly similar result was found in one of the poppies. He concluded that segregation had not yet occurred, and that it must therefore occur in the first pollen division, which separates the tube nucleus from the generative nucleus—the latter then dividing to produce the two sperm nuclei. Strasburger made the suggestion (which we now know to be correct) that the color of the pollen grains is determined by the composition of the plant that produced them, not by the gene content they have just acquired. Correns admitted this possibility but argued, reasonably, that this was a special hypothesis made up to save another hypothesis, and he preferred to avoid such a procedure.

In this same paper, Correns discussed the relation of the genes to the chromosomes. He supposed that the genes were carried by the chromosomes, and he drew a diagram that looks very much like the beads-on-a-string type that later became familiar. One string was labelled *A, B, C,* and so on; on another, closely apposed string *a* was placed opposite *A, b* opposite *B,* and so on. But this was supposed to represent a single mitotic chromosome, which divided in the plane of the paper at each division, except at the time of segregation, when it divided at a right angle to the paper to yield strands *ABC* and *abc.* He also figured what looks now like crossing over, supposing that the pairs of genes could rotate about the long axis of the chromosome, so that *ABc* and *abC* (or *AbC* and *aBc,* and other combinations) could be produced. This was to explain independent assortment ("Mendel's second law").

While this scheme related the genes to the chromosomes, it was wrong in many critical points. There was no explanation of how a single mitotic chromosome came to have maternal and paternal halves; the segregation division was not identified with meiosis; and independent assortment was not related to independent segregation of nonhomologous pairs of chromosomes.

Cannon pointed out in 1902 that there is a close parallelism between Mendelian segregation and chromosome reduction and concluded that this is because the genes are in the chromosomes. Like Correns, he seems to have thought that all the paternal chromosomes went to one pole at meiosis and all the maternal ones to the other; he offered no explanation for independent assortment. De Vries, in 1903, also discussed these questions, and he seems likewise to have supposed that the paternal and maternal chromosomes were separated as groups at meiosis. He accounted for independent assortment by postulating that, at meiosis, the members of individual gene pairs could be freely exchanged between homologous chromosomes (as Correns had supposed) and so would segregate at random.

Guyer, in 1902 and 1903, also understood the situation. The 1902 paper did not mention Mendel, and both were largely concerned with the cytology of sterile hybrids. Guyer did, however, understand that random assortment between different pairs of chromosomes would give independent assortment of genes, although both Wilson and Sutton thought that he had missed this latter point.

Thus there were several people who were close to the correct interpretation at this time, but the first clear and detailed formulation was that of Sutton. W. S. Sutton (1877–1916) was a student of McClung, whose work on the sex chromosomes will be discussed later. Sutton was working with him at the time that McClung first suggested the relation of the X chromosome to sex determination (1901), but Sutton was a graduate student under Wilson at Columbia University when he wrote his two important papers (1902 and 1903). He never finished his graduate work, but did later receive an M.D. degree, and became a practicing surgeon [see biography by McKusick (1960) for more details].

As was pointed out above, the first of Sutton's papers contained the earliest detailed demonstration that the somatic chromosomes (of a grasshopper) occur in definite distinguishably different pairs of like chromosomes. He knew of the earlier work of Montgomery on pairing and of Boveri's paper (also published in 1902) on dispermic eggs. The paper closed with the statement: "I may finally call attention to the probability that the association of paternal and maternal chromosomes in pairs and their subsequent separation during the reducing division ... may constitute the physical basis of the Mendelian law of heredity."

The 1903 paper contains a full elaboration of this hypothesis, including the view that the different pairs of chromosomes orient at random on the meiotic spindles,* thus accounting for the independent segregation of separate pairs of genes seen by Mendel. He suggested, following Fick and Montgomery, that in those cases where both meiotic divisions had been described as longitudinal, the initial pairing had been side by side rather than end to end, as he supposed it to be in insects. The

---

* The cytological demonstration of the random assortment of different pairs of chromosomes was made by Carothers in 1913.

paper contains a discussion and criticism of the 1902 accounts by Cannon and by Guyer, referred to previously.

With this paper, this phase of the history is finished. The conclusions were not at once generally accepted, but they could not be disregarded and stand today as essentially correct. At last, cytology and genetics were brought into intimate relation, and results in each field began to have strong effects on the other.

CHAPTER 6

LINKAGE

With the work of Sutton, the relation of the chromosomes to segregation and to independent assortment became clear. But there was a difficulty, already recognized in 1903 by Sutton and by de Vries: It must be supposed that there are more separately Mendelizing genes than there are chromosomes in the cells. That is to say, there are not enough chromosomes to make it possible to identify each gene with one whole chromosome. De Vries pointed out that this difficulty could be avoided by supposing that genes were freely exchanged between homologous chromosomes at meiosis—the process that he appealed to as an explanation of independent assortment. Boveri made a similar suggestion in 1904. The real solution showed that the principle of independent assortment is not as regularly applicable as was then thought. The discovery and analysis of *linkage,* increasing knowledge of the nature and behavior of the sex chromosomes, and more detailed cytological study of the meiotic prophases finally led to a resolution of the difficulty.

The first report of linkage was that of Correns (1900). He crossed two strains of Matthiola (stocks), one of which had anthocyanin in the petals and seeds, and also had hoary leaves and stems; the other had white flowers and seeds, and smooth leaves and stems. The F_1 had colored flowers and seeds and was hoary. In F_2 he expected to find many types, resulting from independent segregation of three pairs of genes, but actually recovered only the two parental combinations, in the ratio 3 : 1. He suggested that the flower color and seed color might be due to the same pair of genes, but interpreted the relation between color and hoariness as being due to the absence of recombination between two different pairs of genes. He knew of the existence of smooth strains with colored flowers and of hoary strains with white flowers, which confirmed the view that there were two pairs of genes. Later work by Tschermak and by Saunders has shown that the genetic situation is very complex, there being at least four (and probably more) genes in which the various known strains differ. But the most probable interpretation of Correns' original experiment is that he was dealing with two effects of a single gene, and that the different combinations of colors and hoariness that he knew existed are due to mutant genes not present in the strains used in his experiments. This was a degree of genetic complexity unimagined at that time. In any case, he did not recover any recombinations and so thought only of *complete* linkage.

Incomplete linkage was first reported in the sweet pea by Bateson and Punnett (1905), the two gene pairs concerned distinguishing purple flowers from red, and long pollen grains from round ones. The two dominants (purple and long) were contributed by the same parent, and the phenomenon was called "coupling"; the other situation, where one dominant and one recessive of a linked pair come from each parent, was called "repulsion" when it was discovered later by the same authors. Early examples of both types were studied by rearing F_2 cultures, which made the estimation of the frequency of recombination difficult and inexact. Bateson and Punnett concluded that the frequencies found in the early examples fell into a regular series which included a ratio of 7 parental : 1 recombination type and one of 15 : 1. That is to say, the series was supposed to be $(2^n - 1) : 1$, the first member (where $n = 1$) representing independent segregation, with other ratios, such as 3 : 1, 31 : 1, and so on, being expected.* Later work, using test-cross methods on many kinds of plants and animals, has of course shown that there is no tendency for recombination values to fall into any such series; but this supposition led Bateson to formulate the "reduplication" hypothesis that played a large part in later discussions. This hypothesis, though now discredited, must be described.

According to the reduplication hypothesis, segregation does not occur at the time of meiosis but somewhat earlier, and not necessarily at the same time for each pair of genes. The cells that are finally produced, each with a single set of genes, then multiply at different rates to give the observed ratios. It is not easy to see why this scheme was developed, since there is nothing in it that seems related to the $(2^n - 1) : 1$ series, nor is there any independent evidence for the complex and symmetrical pattern of divisions that it requires.† The hypothesis is related to Bateson's

---

* Bateson and Punnett did not present this algebraic formulation nor specifically include the 1 : 1 case in the series, but this seems to be the simplest way of stating their scheme.

† Bateson, an embryologist by training, was impressed by the circumstance that sometimes the two cells arising from the cleavage of a fertilized egg give rise to the right and left sides, respectively, of the embryo. The mirror-image symmetry of these

reluctance to believe that segregation occurs at the meiotic divisions. It was found early that, in some plants (for example, Matthiola, Oenothera) the pollen does not always transmit all the kinds of genes that the eggs of the same individual do. As will appear later (Chapter 10), this is because certain genes prevent the functioning of pollen that contains them—a view to which Bateson was never reconciled. To him these cases were proof of the occurrence of segregation in some division at or before the setting apart of the germinal tissue of the anther—and hence of the inadequacy of the chromosome interpretation of segregation and linkage.

The first suggestion of the relation of a particular character to a particular chromosome was made in 1901, when McClung postulated that the so-called accessory chromosome (now known as the "X chromosome") is male determining. This body was first described by Henking (1891) in the male of the bug Pyrrhocoris. For a long time it was considered doubtful that it was a chromosome, and its uncertain nature and function were the reason for giving it the designation "X." Henking showed that it divides at only one of the meiotic divisions, with the result that it is present in two of the four sperm arising from each primary spermatocyte and absent in the other two. Other investigators (especially Montgomery) confirmed this description for other Hemiptera, and McClung and Sutton found the same relations in several grasshoppers. Sutton, at McClung's suggestion, studied the female; unfortunately the material was difficult and the chromosome number was large, with the result that he counted 22 in the female as compared to the 23 clearly present in the male. Therefore the X was interpreted by McClung as producing maleness, and the supposed significance of the two kinds of sperm was the reverse of the true one.

The correct relation was shown in 1905 for a beetle (Tenebrio) by Stevens; in this case there was also a Y present, smaller than the X, and she showed clearly that the female is XX, the male XY. This result was immediately confirmed by Wilson (also in 1905) for Hemiptera and was soon shown for Orthoptera, Diptera, Homoptera, Myriapoda, and, with less certainty, for various other kinds of animals.

These relations were sometimes interpreted on the basis that the sex chromosomes were not the cause of the differences between males and females, but were merely a kind of secondary sexual character, resulting from some other more basic sex-determining mechanism. The only

---

two halves seemed to him to give a clue as to the nature of heredity—a point to which he returned again and again. This idea seems to have been one source of the reduplication hypothesis.

strong argument in favor of the XY system as the sex-determining mechanism was that it gave a simple way of getting the 1 : 1 sex ratio. But it was known that other situations occur in which fertilized eggs give rise only to females, and unfertilized eggs give rise to either sex. In the group of aphids and phylloxerans, where this occurs, the chromosomes are not too difficult to study, and the work of von Baehr, Stevens, and Morgan soon showed that there were a series of unusual cytological phenomena that constituted a clear confirmation of the XY sex-determining mechanism (see Chapter 13).

Sex-linkage was first reported by Doncaster and Raynor in 1906, in the currant moth (Abraxas); in 1908 Durham and Marryatt demonstrated it in canaries. But in both instances, the results indicated that the female was the heterozygous sex, as was also soon shown in fowl by several workers. Since these results concerned both moths and birds, it seemed that they must be generally applicable; and since the cytological demonstration of the heterozygous nature of the male had likewise been made in many groups of animals, it was also evidently a general condition. This contradiction led to much discussion and speculation, which became pointless with the later discovery of sex-linkage of the type with the male heterozygous by Morgan in 1910 for Drosophila, (and in 1911 for man), and the cytological demonstration of female heterozygosis in moths by Seiler in 1913.

In 1909 there appeared an important paper by Janssens on the cytology of the meiotic divisions in salamanders, especially in Batrachoseps. Janssens raised two questions: Why *two* meiotic divisions both in animals and in plants, when one would appear to suffice for a qualitative chromosome reduction; and how are we to explain the existence of more pairs of genes than the haploid number of chromosomes? He believed that he had found the answers to both of these questions in his chiasmatype theory.

Janssens presented evidence indicating that the longitudinally paired meiotic chromosomes each undergo a longitudinal split, giving a quadripartite structure made up of two daughter strands of each original member of the pair, as others had previously supposed. At the first meiotic division, two strands pass to each daughter cell. Janssens believed that he could show that there had occasionally been an exchange between two of these strands, giving the now-familiar chiasma formation. This accounted for the necessity for two meiotic divisions, since only two of the four strands underwent an exchange at any one level. He also supposed that the two strands involved in an exchange were not, or at least need not always be, sister strands—which meant that there had been an exchange between homologous chromosomes. This in turn "... ouvre le champ à une plus large application cytologique de la théorie de Mendel."

It was at this point that Drosophila entered the scene, so a digression on the history of its use is in order here.

There is a reference in Aristotle to a gnat produced by larvae engendered in the slime of vinegar—this must have been Drosophila. The genus was described and named by Fallén in 1823. It is perhaps to be regretted that his inappropriate name (dew lover) takes precedence over the more descriptive Oinopota (wine drinker), which was used by some early entomologists. The most-studied species, *D. melanogaster,* was described in 1830 by Meigen, and again, under the name *D. ampelophila* (which appears in some of the early genetic literature), by Loew in 1862. This species probably arose in southeastern Asia, but has long been common in all tropical regions; it was introduced into the United States before 1871, probably when bananas began to be imported.

The first person to cultivate Drosophila in the laboratory seems to have been the entomologist, C. W. Woodworth. Through Woodworth, Castle learned of the advantages of the animal; and it was through Castle's work that it became known to other geneticists.[*]

In 1910 Morgan reported the sex-linked inheritance of white eyes in Drosophila, thus resolving the contradiction outlined earlier. Further mutant types were soon found, and one of these (now known as rudimentary) was also found to be sex-linked. Here, then, were two pairs of genes that must be supposed to lie in the X chromosomes, and Morgan saw that one could now test the question causing such wide discussion: Is there recombination between genes that lie in the same pair of chromosomes? The result of crosses between white and rudimentary (1910) showed that recombination did occur, because four types of eggs were produced by females heterozygous for both characteristics. This was a major advance, for it removed the most serious difficulty in the way of accepting the chromosome interpretation of Mendelian inheritance.

It happens that white and rudimentary lie far apart in the X chromosome, with the result that there was no obvious linkage between them; but in the following year, linkage between sex-linked genes was observed by Morgan in several cases—first and most strikingly between yellow body color and white eyes. The possibility that linkage might result from genes lying in the same chromosome had been suggested by Lock in 1906, in his elaboration of de Vries' idea that exchange of mate-

---

[*] A more detailed account of the early laboratory studies, with names and dates, may be found in my biography of Morgan (Sturtevant, 1959).

rials between homologous chromosomes could account for independent segregation; but this had remained merely an interesting suggestion.*

Morgan then applied the chiasmatype hypothesis of Janssens to the results and postulated that linkage is due to the genes concerned lying in the same chromosome pair. The term *crossing over* was introduced, and it was concluded that closely linked genes lie close to each other, more loosely linked ones farther apart. Here, then, in 1911, was the essence of the chromosome interpretation of the phenomena of inheritance. There followed a period of great activity—the usual consequence of a major scientific breakthrough. The next chapter will be concerned with this development.

---

* One of the genes involved in the case described by Lock concerned date of flowering in peas and did not lead to clearly separable classes. While his data indicated linkage, they were not amenable to exact analysis.

CHAPTER 7

THE "FLY ROOM"

When Daniel Coit Gilman became the first president of Johns Hopkins University in 1875, he assembled a remarkable group of scholars to supervise the graduate work there. Among these were two biologists: W. K. Brooks, who had studied with L. Agassiz, and H. Newell Martin, a student of Michael Foster and T. H. Huxley. These two trained a whole generation of outstanding zoologists; among them Edmund Beecher Wilson and Thomas Hunt Morgan were of especial importance in the history of genetics.

E. B. Wilson (1856–1939) took his Ph.D. at Johns Hopkins in 1881, with a thesis on the embryology of the colonial coelenterate Renilla. He then went to Europe, where he worked at Cambridge, at Leipzig under Leuckart, and at Naples. The Naples station, where he returned several times later, greatly influenced him and led to lasting friendships with such men as Dohrn, Herbst, Driesch, and especially Boveri. In 1885 he became the first professor of biology at the newly opened college at Bryn Mawr. In 1891 he became professor of zoology at Columbia University and remained there for the rest of his life. He also spent many summers working at the Marine Biological Laboratory at Woods Hole, Massachusetts. Wilson's early work was largely in embryology, at first descriptive, and later of an experimental nature; his interest here was chiefly in the analysis of the gradual limitation of the potentialities of the cells of the developing embryo and the extent to which "formative stuffs" were involved in development. The first edition of his great *The Cell in Development and Inheritance* appeared in 1896, the second in 1900, and the third (really a new, much larger, and wholly rewritten book) in 1925. This was the standard work for many years and exerted a very great influence on biology. Wilson's own studies on chromosomes began about 1905, with the work on the sex chromosomes referred to in Chapter 6, and led to a series of detailed accounts that are models of accuracy and clarity of expression.

The career of T. H. Morgan (1866–1945) resembled that of Wilson in many respects. He took his Ph.D. at Johns Hopkins in 1890 and then went to Europe, where he was also much influenced by a stay at Naples, and made lasting friendships, especially with Dohrn and Driesch. In 1891 he succeeded Wilson as professor at Bryn Mawr and, in 1904, joined him at Columbia. Like Wilson, he wrote a thesis on embryology, and continued in this field, first with descriptive work, and later with the experimental approach. He also studied the chromosomes in connection with sex determination. At Woods Hole he and Wilson were neighbors, and they and their families were very close friends, both at Woods Hole and in New York.

For all that, the two men were very different. As R. G. Harrison (who was a close friend of both) has expressed it:

> ... Wilhelm Ostwald, in his interesting book on great men of science, classified them, according to their talents, as romantics and classics. . . . To the romantic, ideas come thick and fast; they must find quick expression. His first care is to get a problem off his hands to make room for the next. The classic is more concerned with the perfection of his product, with setting his ideas in the proper relation to each other and to the main body of science. His impulse is to work over his subject so exhaustively and perfectly that no contemporary is able to improve upon it. . . . It is the romantic that revolutionizes, while the classic builds from the ground up.
>
> Wilson is a striking example of the classic, and it is interesting to note that for many years his nearest colleague and closest friend was an equally distinguished romantic.

In 1909, the only time during his twenty-four years at Columbia, Morgan gave the opening lectures in the undergraduate course in beginning zoology. It so happened that C. B. Bridges and I were both in the class. While genetics was not mentioned, we were both attracted to Morgan and were fortunate enough, though both still undergraduates, to be given desks in his laboratory the following year (1910–1911). The possibilities of the genetic study of Drosophila were then just beginning to be apparent; we were at the right place at the right time. The laboratory where we three raised Drosophila for the next seventeen years was familiarly known as "The Fly Room." It was a rather small room (16 by 23 feet), with eight desks crowded into it. Besides the three of us, others were always working there—a steady stream of American and foreign students, doctoral and postdoctoral. One of the most important of these was H. J. Muller, who graduated from Columbia in 1910. He spent the winter of 1911–1912 as a graduate student of physiology at Cornell Medical School and then came back to take a very active part in the Drosophila work.

There was an atmosphere of excitement in the laboratory, and a great deal of discussion and argument about each new result as the work rapidly developed.

In 1909 Castle published diagrams to show the interrelations of genes affecting the color of rabbits. It seems possible now that these diagrams were intended to represent developmental interactions, but they were taken (at Columbia) as an attempt to show the spatial relations in the nucleus. In the latter part of 1911, in conversation with Morgan about this attempt—which we agreed had nothing in its favor—I suddenly realized that the variations in strength of linkage, already attributed by Morgan to differences in the spatial separation of the genes, offered the possibility of determining sequences in the linear dimension of a chromosome. I went home and spent most of the night (to the neglect of my undergraduate homework) in producing the first chromosome map, which included the sex-linked genes y, w, v, m, and r, in the order and approximately the relative spacing that they still appear on the standard maps (Sturtevant, 1913).

The finding of autosomal linkage in Drosophila has been described by Morgan and Bridges (1919) and by Bridges and Morgan (1923) in their accounts of the mutant genes of the second and third chromosomes. The first test of two autosomal genes was made by Sturtevant (February 1912) and showed that black and pink were independent. It was concluded that they were probably in different chromosomes—though this was only a tentative conclusion, since it was known that linkage could approach independent segregation in the frequency of recombination. In March, 1912, Bridges found that the newly discovered mutant curved (wing-shape) when crossed to black, gave no double-mutant types in F_2, so it was clear that autosomal linkage could occur. It was evident that by this time there were more autosomal mutants than there were chromosomes, so Bridges and I began a systematic search by testing the available types against each other. These tests quickly yielded results, but about a week before they did, the second case was discovered by C. J. Lynch, who had made a cross of black to vestigial for another purpose and noted the absence of black vestigial in F_2 (Morgan and Lynch, 1912). This was the first published case of autosomal linkage in Drosophila; it was soon followed by the discovery (Morgan, 1912) that there is no crossing over in the male for these genes. This relation was soon shown to be general for both the second and the third chromosomes. By the middle of July, 1912, the tests carried out by Bridges and Sturtevant had shown that this linkage group (the "second") included not only black, curved, and vestigial, but five additional mutant types. In the same month, we also found two additional types linked to pink, thus beginning the study of the third linkage group (Sturtevant, 1913). The fourth and last linkage group was found by Muller in 1914.

Stevens (1908) described the chromosomes of the female of *Drosophila melanogaster* (under the name *D. ampelophila*) as they are now known, but she found the male difficult to study, and interpreted her figures as meaning that there was a rather small X attached to an autosome, and no Y. This interpretation was followed in early genetic literature on the species, until the work of Bridges (at first on XXY females in 1914), and then of Metz (at first on other species of the genus, also in 1914) established the relations now known. Bridges insisted from the first, and rightly, that the Y is J shaped and longer than the rod-shaped X; but the rest of the group was at first unwilling to accept this, since in other animals (even in other species of Drosophila) the Y was known to be absent, smaller than the X or equal to it, but never larger. A corollary of the earlier interpretation was that there were four, rather than three, pairs of autosomes, with one of them having some sort of relation to the X.

Bridges' cytological work grew out of his studies of nondisjunction. In the first paper on the sex-linkage of white eyes, Morgan reported a few white sons from the original mutant male, which he supposed represented further mutation; there can be no doubt that they were due to nondisjunction.

Further examples kept appearing, and in 1913 Bridges published an extensive genetic analysis of the phenomenon, giving it the name "nondisjunction." Further studies led to no satisfactory causal interpretation until he looked at the chromosomes and saw that females that gave exceptional offspring were XXY in composition (1914). As Bridges understood, this was really a proof of the chromosome theory and made it inconceivable that the relation between genes and chromosomes was merely some kind of accidental parallelism—especially after the publication of his detailed account in 1916, as the first paper in volume 1 of the newly founded journal *Genetics*.

A further consequence of the cytological work of Bridges and of Metz was that it became clear that *D. melanogaster* had three pairs of autosomes—two large and one small—corresponding to the three autosomal linkage groups, of which two were also large and one was small.

By 1915 the work with Drosophila had progressed to the point where the group at Columbia was ready to try to interpret the whole field of

Mendelism in terms of the chromosome theory. The resulting book, *The Mechanism of Mendelian Heredity* (Morgan, Sturtevant, Muller, and Bridges, 1915), is a milestone in the history of the subject.

There had been much reluctance among geneticists to accept the chromosome interpretation. Johannsen, for example, in the 1913 edition of his book, referred to it as "a piece of morphological dialectic"; and Bateson, in a review of *Mechanism* (1916), wrote

> ... it is inconceivable that particles of chromatin or of any other substance, however complex, can possess those powers which must be assigned to our factors [i.e., genes]. ... . The supposition that particles of chromatin, indistinguishable from each other and indeed almost homogeneous under any known test, can by their material nature confer all the properties of life surpasses the range of even the most convinced materialism.

It should be added that by his third edition (1926) Johannsen accepted the chromosome interpretation, and that Bateson thus closed the review (from which the quotation just cited is taken): "... not even the most skeptical of readers can go through the Drosophila work unmoved by a sense of admiration for the zeal and penetration with which it has been conducted, and for the great extension of genetic knowledge to which it has led—greater far than has been made in any one line of work since Mendel's own experiments."

Not all critics were as generous, nor did they always receive soft answers. In short, there were a good many polemical papers; and there surely would have been more if the work had not had the whole-hearted support of Wilson, who had the respect and admiration of all zoologists, making him an invaluable ally.

With the publication of the *Mechanism* and of Bridges' 1916 paper, this part of the story closes. There was still much exciting and fundamental work to be done with Drosophila, and the Columbia laboratory was still the center for such work, but it had become a question of how the chromosome mechanism worked, not of whether it could be demonstrated to be the true mechanism.

There was a give-and-take atmosphere in the fly room. As each new result or new idea came along, it was discussed freely by the group. The published accounts do not always indicate the sources of ideas. It was often not only impossible to say, but was felt to be unimportant, who first had an idea.* A few examples come to mind. The original chromosome

---

* There are, in the later literature, some examples of a concern about priority in the development of ideas in the early period, but at the time such a concern never inhibited free and open discussion.

map made use of a value represented by the number of recombinations divided by the number of parental types as a measure of distance; it was Muller who suggested the simpler and more convenient percentage that the recombinants formed of the whole population. The idea that "cross-over reducers" might be due to inversions of sections was first suggested by Morgan, and this does not appear in my published accounts of the hypothesis. I first suggested to Muller that lethals might be used to give an objective measure of the frequency of mutation. These are isolated examples, but they represent what was going on all the time. I think we came out somewhere near even in this give-and-take, and it certainly accelerated the work.

CHAPTER 8

DEVELOPMENT OF DROSOPHILA WORK

One of the striking things about the early Drosophila results is that the ratios obtained were, by the standards of the time, very poor. With other material it was expected that deviations from the theoretical Mendelian ratios would be small, but with Drosophila such ratios as 3 : 1 or 1 : 1 were rarely closely approximated. This was recognized as being due to considerable differences in the relative mortalities of the various classes in larval and pupal stages, before counts were made. In fact the culture methods were poor, so that these viability differences were very marked, and they still are a source of difficulty.

There has, however, been a steady improvement in the technique of handling the material—most of the early work here being due to Bridges, who brought improved optical equipment and temperature control devices into use, and who had the largest share in the development of improved and standardized culture media.

Bridges was also responsible for the finding and analyzing of a great many new mutants. This undertaking, which was the source of usable material for the studies, was participated in by everyone in the laboratory, but Bridges had the best "eye" for new types and contributed many more of them than did the rest of us. He also had the skill and patience required to build up useful combinations of mutant genes, and many of his multiple strains are still among the most useful, twenty-five years after his death.

Aside from these matters of routine technique, a series of new genetic relations was worked out, and new genetic techniques were developed which made it possible to attack problems that were not recognized or were not approachable before. But one of the first needs was a convenient symbolism; the development of such a system was, in fact, connected with a theoretical question.

51

Mendel used arbitrary letters as gene symbols—*A* and *a*, *B* and *b*, and so on. He started the custom of using a capital letter for the dominant and the corresponding small letter for its recessive allele, a custom that soon became widespread. It was so widespread, in fact, that Cuénot's failure to follow it led to his not being understood.

Bateson introduced the use of mnemonic symbols, the pair being named for the striking characteristic of the dominant allele, for example, *Y* and *y* for yellow and green seeds in peas. Along with this went Bateson's "presence and absence" hypothesis, according to which the recessive was merely the absence of the dominant, that is, the symbol *y* for green seeds indicated merely that the gene *Y* was absent. This hypothesis dominated the field for a good many years, at least until multiple alleles were recognized. There were exceptions, however, such as Johannsen, who was aware that the hypothesis was unnecessary and might be misleading.

Cuénot in 1904 understood the relations between the mouse colors yellow, agouti, and black, which he treated as alleles. However, he gave the genes symbols (*J, jaune; G, gris; N, noir*) which did not suggest this relation, and he seems to have felt that there was nothing unusual or unexpected about the relation. Rather, he felt that the unexpected relation was *lack* of allelism, as in the cases of chocolate, and of pink eyes. By 1907 he listed five "determinants" (= systems of alleles) for colors:

1. *C*, for color in general, and its mutant *A*, for albinism.

2. *M*, for dark eyes and intense coat color and its mutant *E*, for pink eyes and paler coat.

3. "*G* est une déterminant spécial de la teinte du pelage en présence de *C*; il présente un grand nombre de mutations: *G', N,* et *J*."

4. *F* and its mutation *D*, affecting the black pigment (the recessive is now known as chocolate).

5. *U*, for uniform color of whatever hue, and its mutation *P* (panachure) "with a series of variants, p^1, p^2, p^3, ..., p^n, which correspond to varying degrees of spotting."

The last series has not been borne out by later results; but in the case of *G* he had correctly included what are the now-recognized alleles for yellow, white-bellied agouti, agouti, and black, with the order of dominance as now known.

Morgan showed in 1911 that the relation of agouti, black, and yellow was unusual, and the case was interpreted by Sturtevant (1912) as being due to complete linkage between two gene pairs, one for yellow vs. nonyellow, and the other for agouti vs. black. This conclusion was disputed by Little (1912). Then, in 1914, I developed the idea of multiple alleles—basing it largely on the relation between Himalayan and albino rabbits and that between white and eosin eyes in Drosophila, but without realizing that the same interpretation would fit the mouse case. Finally, in 1915, Little pointed out that Cuénot had already given the multiple allele interpretation in papers that were cited and discussed by both Morgan in 1911 and Sturtevant in 1912. Our failure to realize that Cuénot had understood and explained the situation can only be excused by his use of unorthodox symbols and by the fact that he apparently felt there was nothing about the relation that called for elaboration or emphasis.*

With the development of the multiple allele concept, the presence-and-absence hypothesis was abandoned and, at about the same time, the system of gene symbols associated with it broke down for Drosophila. According to that scheme, each pair of genes was named for the somatic effect of the dominant allele, but with the accumulation of many recessive mutant genes, including, for example, a dozen or so for eye-color, it became necessary to name each one for the mutant allele, usually the recessive. The wild type was considered as a standard of reference, usually symbolized as "+." This system, as gradually developed, is now universally used in the Drosophila literature, and essentially the same scheme is applied to microorganisms. The older scheme, or some compromise between it and that used for Drosophila, is still usual for most other higher organisms—a difference in language that sometimes makes for confusion and misunderstanding. It may be added that the Drosophila system itself is now under a good deal of strain, as a result of the phenomenon of pseudoallelism (see Chapter 14).

The first major undertaking after 1913 was the mapping of the new genes as they became available. Here again, while we all took part, it was Bridges who did most of the spadework, and who gradually accumulated and organized the data to produce the maps, which still remain essentially the same in form. With these maps, and with the carefully planned multiple mutant stocks with conveniently located markers, it gradually

---

* It should be added that the view that multiple allelism was to be expected was in accord with the thinking of the time. In 1902 Bateson criticized Mendel's hypothesis of two independently segregating recessives for white flower color in Phaseolus, on the ground that these two postulated recessives should be allelic to each other, and in 1903 he published a note titled "On Mendelian Heredity of Three Characters Allelomorphic to Each Other," the three characters being rose, pea, and single combs in fowl, which are due to two independent pairs of genes. Cuénot's achievement was the recognition of such a case when he found it, and the understanding of its difference from the more usual case of independent pairs of genes.

became possible to work with a precision that was heretofore impossible with any other material.

One of the early discoveries was that of lethal genes. The history of lethals goes back to the work of Cuénot with mice. He reported in 1905 that he had been unable to produce homozygous yellows. On mating yellow × yellow, he obtained 263 yellow : 100 nonyellow, and on testing 81 of these yellows, he found that all were heterozygous. This result led to much discussion by several authors, but in 1910 Castle and Little showed that the ratio of yellow to nonyellow is 2 : 1 (they got 800 : 435, or, adding Cuénot's data, 1063 : 535). There could be no doubt that homozygous yellows were formed, but died before birth. That is, the yellow gene was dominant for coat color and also had a recessive lethal effect.

Baur had already analyzed a similar situation in snapdragons. In 1907 he reported that the form *aurea,* with yellowish-green leaves, when selfed, gives a ratio of 2 yellowish : 1 green, and that the greens breed true, while yellowish × green gives 1 : 1. He postulated that the homozygous yellowish embryos die and, in 1908, showed that most of them germinate but produce seedlings that are almost white. However, they die (evidently from lack of photosynthesis) before counts are ordinarily made. This was the first clear demonstration of a lethal gene.

In 1912, Morgan reported the first sex-linked lethal in Drosophila, which was also the first lethal in which the heterozygote had no detectable phenotypic effect. This gene had no dominant effect, but males carrying it invariably died, giving a 2 : 1 sex ratio; the introduction of marker genes made it possible to locate the lethal on the map of the X. It soon became evident that such recessive lethals constitute the largest single class of mutants in Drosophila; as will appear in Chapter 11, they have been very useful in studies of mutations, and they have also been useful in special techniques for making up and maintaining some complex types of stocks.

The process of crossing over was at once recognized as presenting a mechanical problem capable of experimental study. It was already apparent when the first maps were constructed that one crossing over tends to prevent the occurrence of another one near it. This question of interference was subjected to detailed study by Muller, by Bridges, and by Weinstein. Much accurate information was collected, so that the phenomenon can be described in detail, but no wholly satisfactory interpretation has emerged. It is now known that "negative interference," where a crossover *increases* the probability of another one near it, may occur in the small fourth chromosome of Drosophila, and in certain microorganisms. This matter of interference has again become one of the more actively investigated problems in genetics.

It was soon shown by Bridges that in some regions the frequency of crossing over changes with the age of the female, and by Plough that, in these same regions, the frequency can be influenced by temperature; here again the facts are clear, but their interpretation is not understood.

The chief advance in the understanding of the geometrical relations in crossing over came as a result of the discovery of attached X's. Mrs. Morgan found a very unusual mosaic individual in one of her cultures. As she was examining it, it recovered from anesthetization and flipped off the microscope stage onto the floor. She searched the floor thoroughly, but was unable to find it. Then she reasoned that flies go toward the light when disturbed, so perhaps the mosaic was on the window; there she found and captured it, and was able to recognize it with certainty because of its unusual appearance. The offspring she then obtained from this specimen showed that its ovaries had two X's that always segregated together, and cytological study showed that they were attached at one end, forming a **V** instead of the usual two rods (L. V. Morgan, 1922). This attached-X strain immediately came into very general use as a convenient tool for maintaining sex-linked mutants or combinations in which the females are weak or sterile, and for the rapid multiplication of new sex-linked mutants, or combinations of sex-linked genes.

The two X's in the original attached-X strain were alike, but soon afterward Anderson got a new attached-X line from X-ray experiments, in which the two X's differed in several pairs of genes. Finding that these X's underwent crossing over with each other, he analyzed the results (Anderson, 1925)—an analysis greatly extended with more favorable material later by Beadle and S. Emerson (1933, 1935).

Anderson's study showed that the two X's were attached by their centromere ends, and that genetically these were what had been arbitrarily called their "right" ends. More important, however, was his demonstration that the results could only be accounted for if each chromosome was split lengthwise at the time of crossing over, and if crossing over occurred only between two of the four resulting strands at any one level.

It had been recognized from the beginning of the crossing-over studies that the phenomenon was perhaps one that involved four strands rather than two, that is, perhaps it occurred after the equational split of conjugated chromosomes. This was Janssens' interpretation of the cytological picture, and it was also suggested by Bridges' results on "primary nondisjunction" in 1916. However, there was an alternative explanation possible for Bridges' results, and it seemed simpler to keep to the 2-

strand diagrams as long as they were adequate to explain the data. With these results of Anderson's, and the confirmatory results from triploid females, published at the same time by Bridges and Anderson, the 4-strand interpretation was firmly established.

Crossing over may occasionally occur at mitotic divisions, as was shown by Stern (1936) for Drosophila. Here again it occurs after the chromosomes have undergone a split, and the crossing over is between two (nonsister) strands of the four that are present. The result may be the production of sister cells homozygous for genes that were heterozygous in the original cell.* If these cells then undergo further division and differentiation, there arise adjacent "twin spots" which may show the phenotypes of recessive genes that were originally heterozygous, one in each homologous chromosome.

The production of twin spots was used by Demerec to study the effects of recessive lethal genes that had become homozygous in small areas of a heterozygous individual. He found that some of these are "cell lethal" under these conditions, while others are viable in at least some tissues—either because their normal alleles are not essential for the reactions going on in these tissues or because the necessary substances can somehow be supplied from the rest of the animal.

More recently, it has been found that somatic crossing over occurs regularly in some fungi; this has proved useful for linkage studies, especially in Aspergillus (Pontecorvo and co-workers) and in yeast (Roman and co-workers). Somewhat similar phenomena also occur in bacteria in connection with transformation and transduction and are now under intensive study for their bearing on the mechanism of crossing over.

Bridges followed his genetic and cytological studies of nondisjunction of X with a similar account of nondisjunction of the small IV-chromosomes (1921), and with genetic demonstrations of sectional duplications, deficiencies, and translocations involving the longer chromosomes; these had to wait for the discovery of the properties of the salivary gland chromosomes before they could be fully analyzed (see Chapter 12). Also in 1921 came his studies on triploidy, which will be discussed later (Chapter 13).

Another type of chromosome modification that was worked out step by step may be described here, since it furnished part of the working methods for the analysis of mutation and other phenomena. This is the

---

* That is, somatic segregation occurs. The process is much too rare and too erratic in its incidence to furnish support for Bateson's hypothesis of somatic segregation (Chapter 6), though it sometimes occurs in the germ line and leads to the production of crossover gametes.

inversion of a section of chromosome.

Both Muller and I had earlier found what we considered to be genes that greatly decreased crossing over in the chromosomes in which they were located. We also both found the surprising result that this reduction occurred only when these "genes" were heterozygous; when they were homozygous the reduction disappeared (Sturtevant, 1917).

In 1921 I suggested that there was probably an inversion in the third chromosome of *Drosophila simulans* as compared to that of *D. melanogaster,* and that perhaps this was the nature of the "crossover reducers" in melanogaster. In 1926 Plunkett and I succeeded in demonstrating the existence of the simulans inversion by the locating of more parallel mutants, and, later in the same year, I obtained definite genetic evidence that this was truly the nature of the crossover reducers.

Inversions were soon found to be not rare and were made use of in various ways, especially in keeping sterile or semisterile genes, or combinations of genes. Later they were studied by Sturtevant and Beadle (1936) for the light they throw on the mechanics of crossing over and segregation, and by Sturtevant and Dobzhansky (1936, 1938) in helping to unravel problems in phylogeny. The most important immediate use, however, was by Muller in his studies on mutation (Chapter 11). The full analysis of inversions was not possible until the advent of the salivary gland chromosome technique.

CHAPTER 9

GENETICS OF
CONTINUOUS VARIATION

Galton's use of correlation methods for the analysis of parent-offspring resemblances, resulting in the "Law of Ancestral Inheritance," was described in Chapter 3. This approach was followed by Weldon and by Pearson, the latter, in particular, contributing more sophisticated mathematical techniques, as did Fisher and others more recently.

Karl Pearson was a mathematical physicist by training, and wrote a book, *The Grammar of Science* (1892), that had a great deal of influence. According to Pearson, science is simply description—by which he meant description in quantitative terms—and he was very suspicious of the idea of causality. He was an extreme philosophical idealist; he was also convinced that there is nothing to which the scientific method cannot be applied.

Weldon was a contemporary of Bateson at Cambridge, and the two were at first close friends but later became bitter enemies. According to Bateson's letters, this enmity must have begun about 1890; by 1895 it appeared publicly in a controversy about the origin of the cultivated races of Cineraria, in the course of which each accused the other of deliberately misrepresenting the published statements of an authority that each cited in favor of his own view.

This personal quarrel, which came to involve Pearson as an ally of Weldon, seems to have been a chief reason for the anti-Mendelian stand of both Weldon and Pearson. Bateson felt that they were trying to strangle the new development, and he resisted vigorously. The controversy was pursued in debates and published works for years, and certainly delayed the utilization of the powerful methods of statistics in much of genetics.

In 1902 Yule suggested that the "Law of Ancestral Inheritance" could be thought of as due to the operation of the Mendelian principles in a random-breeding population. In 1904 Pearson disputed this conclusion and attempted to prove that the observed parent-offspring correlations, which had been found in many kinds of organisms, were quantitatively in flat contradiction to the Mendelian scheme. In 1906 Yule showed that Pearson's conclusions rested on the specific assumption of complete dominance for all pairs of genes concerned, and that if dominance was sometimes incomplete, the Mendelian scheme could give correlations throughout the actually observed range. It is now clear that linkage between wholly dominant and wholly recessive genes is a possible alternative interpretation.

In 1905 Darbishire urged that there was no reason why both approaches might not be useful, and both were, in fact, used at about this time by Davenport and others. But the most important influence leading to the general use of statistical methods was that of the Danish botanist Johanssen, beginning in 1903 and culminating in his *Elemente der exakten Erblichkeitslehre* (1909). Johannsen, like Bateson and others, pointed out that the results of the biometrical school were only valid statistically, were of no help in individual families, and gave no insight into the mechanisms involved. But he did recognize the value of statistical methods, and used them extensively. There are twenty-five chapters in his book, and Mendelism does not appear until Chapter 22, the previous ones being concerned almost entirely with the development and use of statistical methods.

Johannsen's work was especially important in emphasizing the distinction between inherited and environmentally produced variations; the words *phenotype* and *genotype* (as well as *gene*) were introduced by him. The distinction was not new. It had, for example, been discussed by Galton under the heading "nature vs. nurture," but it was Johannsen who made it a part of general thought. There had been classifications of variability, as "individual," "fluctuating," "continuous," "discontinuous," and so on, but these were based on the magnitude of the differences rather than on their causes. With Johannsen it became evident that inherited variations could be slight and environmentally produced ones could be large, and that only experiments could distinguish them.

Johannsen's own experiments were largely with beans, and concerned the inheritance of the size of seed in self-fertilized lines. Self-fertilization is the normal method of reproduction here; therefore his plants were in general homozygous—belonging to what he called "pure lines." He found that selection was without effect within such lines, and that two different lines might be only slightly different in size (with much overlapping), but would maintain this slight difference generation after generation. He recognized that the situation was different in cross-fertilizing forms, but was inclined to minimize the effects of selection, and to agree with de Vries that it could produce nothing really new.

In his original paper, Mendel had suggested the beginnings of the theory of multiple genes and had recognized (in connection with flowering time in peas) the confusing effects of environmental differences; but it was not at once evident that continuously varying characters could be studied effectively by Mendelian methods.

In 1902 Bateson pointed out that it should be expected that many genes would influence such a character as stature, since it is so obviously dependent on many diverse and separately varying elements. This point of view was implied by Morgan in 1903 *(Evolution and Adaptation,* p. 277), and by Pearson in 1904. It was developed by Nilsson-Ehle in 1908 and 1909, especially as a result of his experiments with wheat and oats. Nilsson-Ehle applied the idea of numerous separately Mendelizing pairs of genes—plus the confusing effects of environmental differences—to quantitative characters such as size, length of awns, winter-hardiness, and so forth, but his actual detailed analyses concerned somewhat simpler cases, such as color and presence or absence of ligules.

Lock and Castle had both reported crosses in which quantitative differences supposedly gave no increased variability in F_2, but in 1910 East and R. A. Emerson, separately, reported cases in several different plants in which the F_2 was much more variable than the F_1 and interpreted them as being due to the segregation of several pairs of genes. Their joint paper (1913) on maize is a classic in the field and marks the bringing of the inheritance of quantitative characters into the general scheme of Mendelism.

There is another method of studying the inheritance of quantitative characters, namely, the use of selection. This, of course, has a long history; we are here concerned with the relation between it and Mendelian genetics. De Vries was inclined to minimize the effects of selection, and argued that it could produce nothing new. Johanssen had a similar view; but this was so contrary to the point of view of Darwin, Weismann, and the whole generation that followed Darwin, that it was not generally accepted. There followed a series of selection experiments by numerous workers. Those of Johanssen have been discussed, and the circumstance of their being done with self-fertilizing homozygous strains has been pointed out.

The most extensive and most widely discussed selection experiment of the time was that which Castle carried out with rats, beginning in 1914. The "hooded" type here is white with a colored area on the head and usually a narrow, colored line on the back. Castle selected for an increase in the colored area in his "plus" line, and for a decrease in his "minus" line. Selection was effective and ultimately yielded individuals far beyond the limits of the variability of the original series. Both extremes, when crossed to self-colored rats, gave self (that is, uniformly colored) in F_1, and 3 self to 1 hooded in F_2. Castle argued that this pointed to a gradual change in the hooded gene, rather than to an accumulation of modifying genes—a conclusion not accepted by most geneticists. Considerable discussion resulted—some of it (including one paper of mine) rather heated.

Finally Castle crossed the two selected lines, separately, to wild type and examined the grades of the extracted hooded F_2 rats. In each case these were less extreme than their hooded grandparents, that is, more like the original, unselected strain. After two more crosses to wild type and extractions, the two lines were nearly identical, and by 1919 Castle concluded that most of the effects of selection had been due to sorting out and accumulation of modifying genes at loci other than that of the hooded gene, although it seemed probable that a small part of the result was due to minor changes in the hooded gene.

With this result there came to be general agreement that selection operates chiefly through the sorting out of modifiers already present, especially after the demonstration in Drosophila (by Payne, Sturtevant, and others) that such modifiers could be located on chromosome maps by using marker genes. With the realization that modifiers are numerous, it came to be recognized that they are often linked to each other. It is to be expected also that plus modifiers will be linked to minus ones sometimes, thus greatly complicating the analysis. Since practical breeding of domestic animals and cultivated plants depends to a large extent on selection, knowledge about the nature of its action has been of importance in agriculture and horticulture; though perhaps this importance has been more in the understanding of the principles than in the practice of the art of selection.* This art has, however, profited much from the application of statistical methods that grew out of the work discussed in this chapter. Here the work of Fisher has been especially important.

Finally, an understanding of the way selection works has been of the first importance in the application of genetics to the problems of evolution (Chapter 17).

---

* The understanding of the principles may, of course, lead to practical results. As an example, the increased vigor of hybrids was observed long ago by Kölreuter, and even earlier, in the particular case of the mule; but it was the application of Mendelian methods by Shull (1908) and Jones (1917) that led to the most important agricultural uses, beginning with "hybrid corn."

CHAPTER 10

OENOTHERA

As was pointed out in Chapter 3, there was a growing interest in discontinuous variations in the 1890's. In 1901 there appeared the first volume of de Vries' monumental *Die Mutationstheorie,* in which he developed the idea that evolution occurs through discrete steps ("saltations" or "mutations") rather than by gradual changes accumulated by selection. This conclusion rested on a vast amount of data concerning many kinds of plants but was based more especially on the work of de Vries on the evening primrose, *Oenothera Lamarckiana.*

The members of this genus are American in origin, but several of them have escaped from cultivation in Europe, and grow in sandy or disturbed soil there, as they do in much of the United States and Canada. De Vries found a patch of Lamarckiana growing in an abandoned field at Hilversum in Holland, and noticed that two variant types were present. He brought all three types into his garden and found that the typical form produced a series of mutant types, generation after generation. Many of these new types bred true, and most of them differed from the parental form in a whole series of relatively slight respects. It is now known that this is because they differ from the parental form in many genes, and that Lamarckiana is a very unusual and special kind of multiple heterozygote. But to de Vries these new forms were essentially new species, and their sudden occurrence meant that selection had little or nothing to do with the origin of new species that differed from their parents in numerous ways. This was the mutation theory in its original form; it is ironic that few of the original mutations observed by de Vries in Oenothera would now be called mutations.

It seems likely that the properties of these new types were largely responsible for the emphasis placed by many geneticists on the multiplicity of phenotypic effects of single gene changes. It became the custom to emphasize the cases where such multiple effects occur—though surely if geneticists had approached their material without preconceived

62

ideas, the striking thing would have been the relative scarcity of obvious multiple effects of single-gene substitutions.

It soon became evident that the genetic behavior of Oenothera is unusual. The short-styled type (brevistylis) of Lamarckiana was one of the Mendelian characters listed by de Vries in 1900, but it gradually became a puzzle in itself, since nothing else in the plant behaved in so orthodox a way.

The first examples of the new types to be explained were gigas, a tetraploid with 28 chromosomes instead of the usual 14, and lata, a trisomic with 15 chromosomes (Lutz, 1907, 1909). These led to a whole series of observations and experiments with other organisms, but they left unexplained the majority of the Oenothera mutant types, since these were found to have the 14 chromosomes of typical Lamarckiana.

The behavior of these 14-chromosome types when crossed to Lamarckiana, and the results of crosses between various distinct wild forms (biennis, muricata, and so on) were puzzling, sometimes giving "twin" hybrids (that is, two distinct types in F_1 from true-breeding parents), usually giving different results from reciprocal crosses, and usually producing hybrids that bred true. Bateson, and later Davis, suggested that Lamarckiana is really a hybrid—but this suggestion, while probably correct, did little to explain its anomalous behavior. Meanwhile de Vries published many data that seemed to show regularities but resisted all attempts at a systematic analysis.

The solution of the problems was really begun by Renner in a remarkable series of papers that were long disregarded, even by those of us who were actively trying to relate the published data to a scheme consistent with what was known elsewhere. This neglect of Renner's work was undoubtedly due to his use of a system of terminology which was and is very convenient for Oenothera but makes the papers unintelligible unless the special terminology is first learned. When it is learned, the papers are found to be written in a very clear and logical style.

The series of papers began in 1913 with one on fertilization and early embryology; it showed that a suggestion of Goldschmidt's (merogony) was incorrect. This was followed in 1914 and 1917 by a study of the embryos and seeds from Lamarckiana after self-pollination and after crossing with other species. These studies showed that Lamarckiana is a permanent heterozygote between two "complexes" called "gaudens" and "velans." After self-fertilization about half of the seeds contain inviable embryos. Half of these die at an early stage, and the other half at a later one. Renner concluded that these inviable seeds represented the gaudens-

gaudens, and velans-velans types, respectively, whereas the viable seeds were all gaudens-velans heterozygotes. In agreement with this was the fact that crosses (for example, to muricata) that gave twin hybrids gave fully formed viable embryos in almost all the F_1 seeds. Renner here developed the hypothesis of balanced lethals, though he did not use that term. He also suggested, especially in the 1917 paper, that such "mutant" types as nanella and rubrinervis arise from recombination between the two complexes.

In biennis, muricata, and suaveolens, the functional pollen is all of one kind, and the eggs are mostly of a different kind, so that crosses with these species yield reciprocal hybrids that are different. Renner studied the pollen in these species and in hybrids from them and showed (1919) that each produces two kinds of pollen in equal numbers, which are distinguishable especially by the shape of the starch grains they carry. Only one of these types is functional, as shown by the shapes of the starch grains in the pollen tubes in the styles, and in the pollen of their hybrids. Here was a direct demonstration of a pollen lethal and also a clear disproof of the idea of somatic segregation that Bateson continued to insist on in certain cases in Matthiola and in Pelargonium, where the pollen also fails to transmit some of the genes that may be recovered from eggs of the same individual.

The eggs, especially in muricata, only rarely transmit the complex that is transmitted by all the pollen, and Renner's studies (1921) showed why this occurs. In Hookeri, which is homozygous, or in Lamarckiana, where the eggs are of two kinds in nearly equal numbers, he found that the uppermost (micropylar) megaspore of the four that result from meiosis is regularly the one that functions to produce the gametophyte. That is, it has an advantage due to its position. But in muricata the upper megaspore functions in only about half of the ovules; in the other half the basal of the four is functional. Evidently the "rigens" complex has an inherent advantage over the "curvans" one that usually enables it to function even when it occupies the less favorable position—although it never functions in the pollen. Here then, by study of the nature of the cells themselves, Renner succeeded in solving the problem of how the Oenothera species maintain their balanced condition—both the "homogametic" condition of Lamarckiana (where eggs and sperm both transmit both complexes) and the "heterogametic" one of muricata and similar forms (based on pollen lethals and megaspore competition).

These results left unexplained the nature of the "complexes," which Renner interpreted as groups of linked genes, and he set about analyzing them in terms of separable components. It was soon apparent that the linkages are not constant. The dominant gene for red midribs on the leaves is completely linked to the complexes in muricata and biennis, but segregates independently of them in Lamarckiana. The various hybrids show one or the other of these kinds of behavior, but almost never an intermediate type with moderate linkage. In some of the hybrids, such as curvans-velans (from muricata by Lamarckiana), there is rather extensive recombination between the complexes, and Renner made use of such hybrids to dissect the complexes into their component parts. The most extensive account of these studies appeared in 1925. In this paper Renner concluded that if two genes are independent in any combination, they are in different pairs of chromosomes, and if these same two are closely linked in another combination, then in the second case the two pairs of chromosomes are not showing recombination. He suggested that the explanation was probably to be sought in the chromosome rings that Cleland had already described in Oenothera.

Cleland reported in 1922 that the 14 chromosomes of *Oenothera franciscana* do not form 7 bivalents at meiosis, but 5 bivalents and a ring of four. In 1923 he recorded still larger rings, including a ring of 8 and one of 6 in biennis, and a ring of 14 in muricata. He showed that alternate chromosomes in these rings pass to the same pole at the first meiotic division and suggested that this behavior might be related to the frequent linkage of characters that occurs in Oenothera.

Similar chromosome rings were observed in Datura by Belling, who in 1927 suggested that they were due to the past occurrence of translocations, so that two original nonhomologous chromosomes, with ends that may be represented as a.b and c.d, gave rise to two new chromosomes that between them carried the same genes but had the arrangement a.d b.c (or a.c b.d). He specifically suggested that the repeated occurrence of such translocations might give rise to the large rings of Oenothera. This suggestion was then followed up by Cleland and Blakeslee (1930) and by S. Emerson and Sturtevant (1931), who showed that it could be utilized to give a self-consistent scheme for the numerous configurations known, and that this scheme was also consistent with the variable linkage reported by Renner. With this result, the peculiar genetic behavior of Oenothera was at last brought into line with the general Mendelian scheme.

More recently these principles have been used by Cleland to build up a very extensive series of analyses of the chromosome makeup of a large number of strains collected over most of the United States, and by Renner and others to locate particular genes in particular chromosomes. The discovery of a "V-type" position effect in Oenothera by Catcheside will be referred to later (Chapter 14); the most recent advance in the unravelling of the genetic complications of the group is the discovery by Steiner (1956) that the egg complexes of many wild forms of the eastern United States carry self-sterility alleles of the oppositional type already known in the remotely related *O. organensis* (S. Emerson, 1938).

CHAPTER 11

MUTATION

Early studies on variation were concerned with phenomena that were in part due to recombination, and the origin of new types could be studied effectively only when the question could be formulated in terms of the origin of new genes. The first question was: Do new genes in fact arise, or is all genetic variability due to recombination of preexisting genes? This question was seriously discussed—though the alternative to mutation seems to be an initial divine creation of all existing genes.

It is true that most strains studied are of hybrid derivation—or may be so—and that recessives may remain undetected for many generations. The result is that, while one may argue that the origin of new genes is a logical necessity, it is not so easy to specify that a given gene must be of recent origin. Perhaps the clearest evidence in the early studies came from the sweet pea, which was known to be descended entirely from a single small collection of wild material (from Sicily) and to have been propagated for many generations, mostly by self-pollination, before new types appeared. The history of the various color types was known, and left no doubt that particular genes had arisen suddenly in cultivated material. But most of the newly arisen forms in such cases are recessive to the original type, and, on the long-accepted "presence and absence" hypothesis, represent merely losses of genes—not the acquisition of new ones. As late as 1914, Bateson discussed the possibility that evolution had come about solely through the loss of genes, followed by recombination.

In diploid organisms there are only two classes of genes whose mutability can be studied simply and effectively—newly arising dominants, and sex-linked genes, where in the heterozygous sex each individual is effectively haploid. Both were used in early work, and special methods were developed fairly quickly for the study of autosomal recessives, especially in self-fertilizing plants.

In the 1890's de Vries studied the genetic behavior of variegation in the flowers of Antirrhinum, the snapdragon. Here the flowers are white or yellow, with red stripes, the variegation being recessive to the self red condition. On a variegated plant, the size of the red areas is variable, and sometimes includes an entire branch. De Vries showed (1901, *Die Mutationstheorie)* that the flowers on these red branches behave like the red flowers on a wholly red F_1, that is, on self-pollination they give 3 red : 1 variegated. The results were not given a Mendelian interpretation because, at the time the experiments were done, de Vries did not know about Mendelism. He did, however, suggest that some kind of segregation *(Spaltung)* was occurring, so that self color was somehow split off from the variegated element; he knew, and emphasized the point, that no true whites were produced.

Correns (1909, 1910) studied similar cases in the four o'clock, Mirabilis, the most detailed account concerning variegated leaves. He gave a Mendelian interpretation but thought of segregation rather than of mutation—though, like de Vries, he found that the change was in one direction only: from variegated (pale green and dark green) to self dark green. His main emphasis was on the change in somatic tissues from a homozygous variegated condition to a heterozygous (self/variegated) one. Neither de Vries nor Correns seems to have been clear that the variegation itself, within individual flowers or leaves, could most easily be interpreted as being due to the occurrence, late in development, of the same change responsible for large sectors, which could then be transmitted. This view, and the interpretation of the whole series of events as recurrent gene mutations, was developed by R. A. Emerson (1914) as a result of his studies on the variegated pericarp of the "calico" type of maize. He also pointed out what was evident in his results and in those of de Vries and of Correns (when finally mutation was thought to cause variegation), namely, that the mutation regularly occurred in only one of the two genes in a single homozygous cell.

The regular and frequent occurrence of mutations in such cases was made clear by these papers, but there was, and still remains, a reluctance to generalize from them about the mutation process, since the usual frequency is so very much less than they report—and is for that reason much more difficult to study.

The early work on Drosophila, especially that of Bridges, furnished several instances of the occurrence of new dominant genes, and many new sex-linked recessives, in pedigreed material where the event could be analyzed in some detail. These examples confirmed the conclusion that mutation occurs in a single gene in a single cell, and that it can occur at any stage of development. In fact, Muller concluded that the frequency per cell is probably the same for each stage in the germ line.

These results were merely qualitative, since the frequencies were too low for a quantitative study and were also strongly influenced by the personal equation of the observer. What was needed was an objective index, and one that would recognize a class of mutations that was frequent enough to give significant numerical values. Both of these requirements were met in the system devised by Muller for the study of newly arisen sex-linked lethals in Drosophila.

As first used by Muller and Altenburg (1919), the method depended on the study of the sex-ratios from individual females that were heterozygous for sex-linked "marker" genes. This technique made possible an objective and unambiguous determination of the frequency of sex-linked lethals, but it was very laborious, since rather detailed counts had to be made for each tested chromosome. There was a suggestion of an increase in the rate of occurrence of new lethals with an increase in temperature; this was later confirmed with improved techniques. It was also evident that the rate per unit time could not be as high in man as in Drosophila, for the approximate 1 : 1 human sex-ratio would then be quite impossible.

Muller then improved the technique by using the "ClB" chromosome that he found in his experiments. This is an X chromosome that carries a crossover reducer (later shown to be an inversion), a lethal, and the dominant mutant gene Bar. As is well known now, this chromosome made it possible to detect new sex-linked lethals without anesthetizing the flies or making counts—merely by rapid examination of individual culture bottles. In this way one could test many more chromosomes than possible ever before and thus could obtain adequate data on lethal frequencies.

The first great advance through the use of this technique was the demonstration of the mutagenic action of X rays (Muller, 1927).

X rays were reported in 1895 by Roentgen, and, in the next year, radiation burns caused by them were reported by L. G. Stevens. There followed a rapidly increasing volume of literature on the biological effects of X rays and (beginning in 1900) of radium rays. "Radiation sickness," described in 1897 by Seguy and Quenisset, and by Walsh, was characterized by heart symptoms, headaches, and insomnia. The first successful therapeutic use seems to have been by Steenbeck, who in 1900 reported the destruction of a small tumor in the nose of a patient. Radium burns were reported in this same year by Walkoff and by Griesel—a result confirmed in 1901 by Becquerel and Curie, who deliberately exposed themselves. In 1901 Rollins reported that heavy doses to guinea pigs were lethal without the occurrence of any visible surface changes.

The carcinogenic properties were reported in 1902 by Frieben, who described a carcinoma of the skin following X-ray exposure. In 1903 Bohn concluded, as a result of studies on the gametes and the fertilized eggs of sea urchins, that the primary damage caused by radium treatment was in the chromatin. In 1904 Perthes reported on the cytological effects of radium on the developing eggs of Ascaris and suggested that the chromosomes were fragmented—though he wondered if this result might be due to accidental cutting by the microtome knife in making his preparations. In 1905 Koernicke treated Lilium with radium and concluded that there was a true fragmentation of the chromosomes.

Numerous attempts were made to induce mutations by high-energy radiations and also by other physical and chemical treatments. Mac-Dougall, Vail, and Shull (1907) treated Oenothera with radium, and Morgan (1911) and Loeb and Bancroft (1911) used it with Drosophila. In all these cases, a few mutants occurred, but at the time it seemed most probable that they were not produced by the treatment, since mutations also occurred in the controls. It is now clear that the genetic techniques then used were not adequate for the demonstration of an increase in mutation frequency of the magnitude likely to occur.

Genetic effects were found, however, in Drosophila by Mavor. In 1921 and 1922 he showed that X rays produce a marked increase in the frequency of nondisjunction, and in 1923 he found that they also affect the frequency of crossing over. Anderson confirmed the nondisjunction effect and in 1925 showed that one of the exceptional females produced had her two X's attached—this being the first induced chromosome rearrangement.

When Muller applied his "ClB" technique to the study of sperm treated with X rays, it was apparent that there was a marked increase in the frequency of newly arising lethals. This paper (1927) marks the first clear example of the artificial induction of mutations.

Stadler had begun his studies with barley at about the same time that Muller's work with X rays began, but since he was using an annual plant, his results were not available until after Muller's were published. His first 1928 paper did furnish independent confirmation of Muller's result, on different material studied with a different technique.

Stadler irradiated seeds of barley. At the time of treatment, the embryo was already formed, and each of the several main shoots of the developing plant was represented by a single diploid cell. When the plant was mature, each shoot was separately self-pollinated and the seeds were planted. Any induced recessive segregated in the ratio 3 : 1 and could be identified as induced, since it would not appear from other shoots of the same plant. In all, Stadler found forty-eight mutations in distinct seedling characters from irradiated seeds, and none from a large control series. Some of these seeds had been given X rays, others had been exposed to radium—the latter proving to be also mutagenic.

These results of Muller and of Stadler at last opened the way to an experimental attack on problems having to do with mutation, and they and others began an extensive study of these problems—a study that is still being very actively prosecuted.

Muller, in his first account, concluded that high-energy radiation is dangerous not only to the exposed individual, but to his descendants as well. This conclusion has come to be generally understood and accepted, and is now of much public concern in connection with the medical and dental use of X rays, and the worldwide distribution of radioactive isotopes produced by nuclear fission.

In this same paper, Muller made two tentative suggestions that were soon shown, by Muller himself and by others, to be incorrect: first, that "spontaneous" mutations might be largely or wholly due to normal background radiation, and second, that the relation between dosage and mutation rate might be exponential rather than linear.

Stadler, in his second paper in 1928, reported (in an abstract about half a page long) three major discoveries made with barley:

1. Seeds soaked in water to initiate germination gave almost eight times the mutation rate of dormant seeds.

2. Mutation rate was independent of the temperature at the time of irradiation.

3. The relation of mutation rate to total dosage was linear—doubling the dose doubled the mutation rate.

In 1929, Hanson and Heys studied the induction of lethals in Drosophila by radium. They interposed lead shields of different thicknesses and recorded the ionization (using an ionization chamber) in each treatment. The curves for ionization and for mutation rate were superposable, leading to the conclusion that ionization is responsible for the mutations and that the relation is a simple, direct one—a "one-hit" phenomenon. In 1930, Oliver confirmed this conclusion by showing that varying the dose by varying the duration of exposure to a constant X ray source also gave a linear curve relating dosage to mutation rate. It was apparent that projecting this curve to a zero dosage did not give zero mutations, and Muller and Mott-Smith showed in this same year that the amount of background radiation is far less than would be required to produce the normal "spontaneous" frequency of mutations.

It came to be very generally accepted that total ionization is all that need be considered in connection with radiation-induced mutations, at least within a strain. Stadler's dormancy results should have suggested that it was dangerous to argue from sperm (which are essentially dormant) to actively growing tissues; but it was not until rather recently that it was shown (by Lüning, Russell, and others) that there are differences in dose response in different tissues and developmental stages in Drosophila and in mice. These studies, now being actively prosecuted, fall outside the scope of this book.

The induction of mutations by chemical means was attempted by many people over a long period, but until 1941 there were no clear and convincing positive results. In that year Auerbach and Robson obtained clear evidence that mustard gas is mutagenic. They were led to study it because pathologists had found much similarity between the appearance and behavior of skin burns induced by mustard gas and by irradiation. The mutation effect was observed in a (British) government-supported war project, and the information was classified and could not be published until 1946. With its publication began the present active and productive study of chemical mutagens—another development outside the scope of this book.

CHAPTER 12

CYTOLOGICAL MAPS AND THE CYTOLOGY OF CROSSING OVER

Rearrangements of the chromosomal material were first detected in Drosophila by genetic methods. Deficiencies were reported by Bridges (1917) and by Mohr (1919), duplication by Bridges (1919), translocation by Bridges (1923), and inversion by Sturtevant (1926). These were all of spontaneous occurrence, and none of them were cytologically identifiable by the methods then available.

The first successful cytological attempt to analyze the chromosomes from a genetic point of view, in terms of units less than a whole chromosome, was made by Belling. By 1924, Belling and Blakeslee reported a series of extra-chromosome types in Datura. In this paper they developed the idea that the meiotic pairing of chromosomes, even at a late stage (diakinesis), could be used to determine the homologies of separate arms. They described "secondary trisomics," in which the extra chromosome was made up of two like arms of a normal chromosome, presumably having arisen by a somatic division that was normal except that the centromere had divided transversely to the long axis of the chromosome instead of parallel to that axis. These types gave evidence as to the phenotypic effects of the separate arms. Unfortunately, however, there were few mutant genes available in the plant, and the more critical earlier stages (pachytene) of meiosis were not favorable for study in Datura.

A different attempt to get cytological information on the genetic composition of individual chromosomes was developed for maize by McClintock, Randolph, Longley, and others. Here it was possible to study the pairing of homologous chromosomes at pachytene, when they were longer and showed more recognizable detail than at the later stages

73

chiefly studied by Belling. A large series of mutant genes was known, with linkage maps well understood, and a series of chromosome rearrangements was collected. This was the most hopeful material for a detailed correlation between cytologically visible structures and linkage maps—until the development of the salivary gland chromosome technique for Drosophila in 1933 (see p. 75).

In Drosophila, the first evidence relating a cytologically visible structure to a linkage map was Anderson's proof (1925) that the centromere end of the X is the right end of the linkage map—as pointed out in Chapter 8.

Muller's original report on the mutagenic effects of X rays (1927) stated that he had also recovered "a high proportion of changes in the linear order of the genes." The presence of rearrangements was confirmed by cytological studies (Painter and Muller, 1929). They showed, especially by a study of long deletions, that the genetical map of the X, based on crossing-over frequencies, does not correspond with the intervals measured on the metaphase chromosomes—although the sequential order is mutually consistent. By 1931 they showed that a large section of the right (centromere) end of the X is "inert" (contains very few genes) and in 1930 produced a "cytological map" of the X.

Dobzhansky (1929, 1930) studied X-ray-induced translocations involving the second and the third chromosomes with the small fourth. The positions of the break-points were determined cytologically, and also genetically, by determining the apparent locus of a fourth-chromosome gene (eyeless) on the maps of the longer chromosomes. There resulted cytological maps of these chromosomes, which were consistent with the sequences of loci on the genetical maps but showed that, as Muller and Painter found for the X, the intervals were not proportional. That is, there were relatively long sections with relatively little crossing over, and relatively short ones with much crossing over.

In 1931 two papers that appeared independently demonstrated that recombinants arising from genetic crossing over are accompanied by exchange of cytologically visible markers. The first of these, by Creighton and McClintock, utilized a translocation and a "knob" (heterochromatic end) in maize; the second, by Stem, utilized an X of Drosophila with an arm of Y attached to its right end, and an X-IV translocation. In both cases, two marker genes between the cytologically identifiable regions were available, and it was demonstrated that recombination between the marker genes was regularly accompanied by recombination between the cytological markers. These papers gave the final cytological proof that genetic crossing over is accompanied by an exchange of parts between chromosomes.

The metaphase chromosomes of Drosophila are very small and show little structural detail. The use of brain cells, introduced by Frolowa (1926), gave somewhat larger figures than the previously studied oögonial cells, but the breakpoints were still only approximately identifiable. No cytological analysis was possible for short deficiencies or duplications, for inversions within a single chromosome arm, or for translocations involving exchanges of nearly equal parts.

This was radically changed with the advent of the salivary gland chromosome analysis. The existence of large, banded strands in the salivary gland nuclei of Chironomus larvae was recorded by Balbiani in 1881, and this condition in the salivary glands, Malpighian tubes, and in some cells of the gut of several groups of Diptera was studied by several authors after that date. The condition was observed in living, intact larvae and was also studied in fixed and stained sections. The usual interpretation was that these strands formed a continuous spireme, with only two free ends. Only in 1933 (January) was this shown to be incorrect, when Heitz and Bauer studied the Malpighian tube cells of Bibio by the squash technique instead of sections. Pressure spread the threads, and they were able to show that there was a definite number of distinct worm-like bodies, tangled in an unanalyzable mass in the living cells or in sectioned material, but separate and countable in their squash preparations. They further found that the number of bodies was the haploid one, and that the relative sizes were like those of the metaphase chromosomes. They concluded that each of the worm-like bodies was a closely conjugated pair of homologous chromosomes,* and they also pointed out that each of them had a characteristic banding pattern and characteristic ends, recognizable from cell to cell and from one larva to another one.

Heitz had previously (1928) shown that in the liverwort Pellia there are heterochromatic regions in the chromosomes at somatic divisions, and that these tend to aggregate into a common chromocenter in resting stages. In 1933 (December) he showed that similar relations are to be found in Drosophila—specifically, that much of the basal region of the X is heterochromatic (a result which he correlated with the "inert" region of Muller and Painter), and that there is a common chromocenter to which the salivary gland chromosomes are attached. However, he found the salivaries difficult to study and did not carry his analysis very far.

---

* It had already been shown, by Stevens and by Metz, that homologous chromosomes of Diptera usually show "somatic pairing" at ordinary somatic divisions.

In the same month (December, 1933) there appeared Painter's account of the salivary gland chromosomes of *Drosophila melanogaster,* which he showed were quite workable, if one studied old larvae nearly ready to pupate. In this paper he presented a drawing of the euchromatic part of the X, with over 150 bands, and with 13 corresponding points shown, that had been determined both cytologically and genetically from a long deletion, seven translocations, and two inversions (one of the latter being the familiar ClB). Here at last was a detailed correspondence in sequence between the crossover map and cytologically visible landmarks, and a technique that was clearly capable of refinement to give the precise loci of genes in terms of recognizable bands. Instead of two or three landmarks per chromosome (the ends and centromeres), there were now hundreds, and there soon came to be thousands for the whole complex.

There followed a series of studies in several laboratories, which rapidly gave more and more detailed cytological maps of all the chromosomes, both of melanogaster and of other species. In the case of melanogaster, where the available genetic data were much more extensive, the detailed studies of Bridges were especially useful, and his drawings of the salivary gland chromosomes of that species (1935, 1938, 1939) are still the standards.

In 1935 Bridges recognized 725 bands for the X chromosome, 1320 for the second, 1450 for the third, and 45 for the fourth. In 1938 the number for the X was raised to 1024, and in 1939, the number for the right limb of the second was raised from 660 to 1136. He recognized that even these numbers did not exhaust the potential resolving power of the method. Bridges also developed a convenient system for designating each band—a system that is still in use.

An early result of the salivary gland studies was the discovery of *repeats* by Bridges (1935) in Drosophila and by Metz (1938) in Sciara. These repeats were shown to be either "direct" or "reverse" in their orientation with respect to each other, and to be either adjacent or separated by other regions. Their origin is not altogether clear, but their frequent occurrence is, as Bridges pointed out, of considerable evolutionary interest, since they furnish extra genes that are presumably not needed by the organism, and that may be of importance in making possible the origin of new genes with new functions.

Another result of these studies, recently found to be of great interest, is that of the "puffing" of certain regions. It was shown by Metz (1938) that certain regions of the salivary gland chromosomes undergo a reversible process in which the bands swell and show a much looser structure. Pavan (1952) found that in Rhynchosciara this is a regular phenomenon, particular bands undergoing puffing at specific developmental stages.

This has been fully confirmed by Rudkin, Beermann, and others, and the subject is currently being actively studied—especially by Beermann and his co-workers—because of its bearing on questions relating to the timing of gene action in development.

The original interpretation of chiasmata by Janssens (1909) rested on the assumption that the initial separation of the four strands involved in a tetrad was always, at every level, such that two sister chromatids remained together in each of the separating areas. On this basis there is a one-to-one correspondence between a visible chiasma and a genetic exchange; at such a visible chiasma two of the four chromatids have undergone a crossing over, and these two are nonsister chromatids.

This assumption was not proven then, as was soon pointed out by Robertson (1916) and others—and it has still not been proven. It may be that, at some levels in a tetrad, the initial separation does not separate one pair of sisters from the other (reductional separation), as Janssens supposed, but is equational, separating two nonsisters from the other two (also nonsister) chromatids. Otherwise stated, it may be that if one visualizes the four chromatids as straight untwisted rods, the initial two-by-two separation may occur in either of the two geometrically possible planes. If this assumption is accepted, then there is no necessary relation between visible chiasmata and genetic crossing over. As Wilson put it in 1925:

> To the author, all seems to point to the conclusion that the mechanism of crossing-over must be sought in the pachytene stage during the period following synapsis. . . . The genetic evidence . . . leads almost inevitably to the conclusion that crossing-over must involve some process of torsion and subsequent splitting apart . . . but we must admit that on its cytological side the problem still remains unsolved.

The cytological study of the meiotic process was actively prosecuted at about this period, in an attempt to see what really happened at crossing over. Among the numerous workers then, perhaps the most important was Belling, who studied more especially plants of the lily and related families. From 1926 to 1931 he published several "working hypotheses," based on the assumption of random breaking of the thin, paired chromosome strands, with reunion of the broken ends, which could lead to interchanges between homologues if two breaks happened to occur at the same level. In the later forms of these models, he related the phenomenon to the production of new daughter chromatids—an idea that has been involved in many of the more recent interpretations.

The most ambitious attempt at a general scheme is that of Darlington, embodied in a long series of papers and first developed in detail in his book of 1932. This scheme was very generally accepted, and for a time came to be considered the very backbone of cytogenetics. It depends on Darlington's "precocity theory," which he sums up as follows: "Meiosis differs from mitosis in the nucleus entering prophase before the chromosomes divide instead of after they divide."

According to this scheme there is a tendency for chromosomes and their constituent parts to form pairs of like elements at the beginning of prophase. If chromosome division has already occurred (as at mitosis), this affinity is satisfied by the fact that daughter chromatids are still closely apposed; at meiosis it leads to a conjugation between homologues. In the latter case, when the conjugated chromosomes divide there are four apposed strands, and the attractive force is supposed to be satisfied when two elements are apposed. Therefore there occurs a separation (reductional) into two double bodies, each made up of a pair of sister elements. If, now, there has been an exchange (that is, a crossover), there will be a chiasma corresponding to it, since only in this way can each part undergo a reductional separation. These chiasmata hold the structure together and ensure that the orientation at the metaphase of the first meiotic division will lead to the passage of two chromatids to each pole.

This scheme was elaborated in great detail, and gave a satisfying geometrical picture, which was correlated with the genetic results by many workers. To many of us, it came to be accepted as basic (see, for example, Sturtevant and Beadle, 1939). But there were skeptics from the beginning. Belling was very critical of much of the scheme, as were Sax and others. It was soon apparent that, in some forms, the chromosomes are visibly double at the time they first conjugate; the view that the initial separation is always reductional at each level was questioned as being an unsupported hypothesis. It was pointed out that quite regular first meiotic segregation occurs without any accompanying crossing over in the Drosophila male and, sometimes, also in the female. Some of the supporting observations themselves were questioned—notably the quantitative agreement between observed frequencies of crossing over and counted numbers of chiasmata. Here the fact is that the counting of chiasmata can be carried out in a really convincing and unambiguous manner in only a few very favorable objects, and these unfortunately do not include any forms in which there is a considerable body of evidence on the total frequency of crossing over.

Much of the critical discussion in this field is too recent for inclusion here; it is based in part on the suggestion that crossing over may occur much earlier than the detailed side-by-side pairing at synapsis, through chance overlapping of the very thin (and mostly unpaired) threads—in which case the genetically important event occurs before cytologists normally begin looking (Taylor, Grell, and others). It is also probable that a final scheme will depend in large part on the results obtained with bacteria and with bacteriophage, which cannot yet be fully evaluated in comparison with the chromosomes of higher forms.*

---

* I should like here to enter a protest against the current use, especially by students of bacteria and bacteriophage, of the word *chromosome* as synonymous with *linkage group*. A chromosome is a body that is visible under the light microscope, contains both DNA and other material, and has a whole series of reasonably well-understood properties. The bodies so designated in bacteria and in bacteriophage are very much smaller, seem to be wholly DNA, and lack many of the properties of true chromosomes. They do agree in containing the genes and in being subject to recombination. No one can question the importance of the studies being made about them—but it seems essential to avoid confusion by using a different term; *genophore,* suggested by Ris, seems appropriate and desirable.

CHAPTER 13

SEX DETERMINATION

Theories of the determination of sex were already numerous in Aristotle's time, and he discussed many of them. His own view was that there is, in each embryo, a sort of contest between the male and female potentialities, and the question of which prevails, that is, the frequencies of the two sexes, may be influenced by many factors, such as the age of the parents, the direction of the wind, and so forth. This idea of a competition between opposing influences has been involved in most theories, down to the present; the current form is described by the term *genic balance*.

The existence of males and females, in approximately equal numbers, continued to intrigue both philosophers and biologists. Thomson in 1908 wrote: "The number of speculations as to the nature of sex has been well-nigh doubled since Drelincourt, in the eighteenth century, brought together 262 'groundless hypotheses,' and since Blumenbach caustically remarked that nothing was more certain than that Drelincourt's own theory formed the 263rd. Subsequent investigators have long ago added Blumenbach's theory to the list."

The discovery of the sex chromosomes and the demonstration of their relation to sex determination, have been described in Chapters 6 and 7. One result of the nondisjunction studies was not pointed out: the X-bearing sperm is not in itself female determining, since it may produce a male if the egg carries no X, that is, sex is determined by the composition of the zygote. This conclusion was confirmed and extended by studies of gynandromorphs. Morgan and Bridges (1919) showed, by a detailed study of a large series of these, that the separate parts of the body of Drosophila are largely independent in their determination, and that the sex of each part is due to its chromosome composition.

Another relation established by Bridges in his nondisjunction work is that diploid individuals of Drosophila with the composition XXY are normal and fertile females, and that those with a single X and no Y (XO) are normal males in appearance—though they are sterile. That is to say, the Y is not the primary sex-determining agent.

A much fuller analysis resulted from the study of triploids by Bridges (1921). These results were based on a remarkable series of parallel genetical and cytological studies, and furnished convincing proof that sex in Drosophila is due to a balance between the number of X's (which have a net female-producing effect) and the number of sets of autosomes (which have a net male-producing effect). Thus, addition of an X to the normal male composition produces a female, while addition of a set of autosomes to the female composition produces an intersex. This conclusion has been fully confirmed by the later finding of a few additional types from the tetraploid females.

Bridges interpreted these results on the basis of the "genic balance" idea that he had deduced early in 1921 from his studies on haplo-IV individuals, which have a recognizable phenotype that differs from the wild type in a number of respects. Bridges pointed out that there is evidence for the existence of numerous genes affecting a given character—some in one way, some in another. Each individual represents the resultant of a particular balance between these variously acting genes. It is unlikely that any given chromosome, or section of a chromosome, will have a set of genes with the same net effect as the whole complement; therefore it is to be expected that duplications or deficiencies will cause changes in this balance and will alter the phenotype—usually to the detriment of the individual, and often with even a lethal effect.

In the case of sex, this interpretation means that there are genes with male-producing effects, and others with female-producing ones; and that the former predominate in the autosomes, the latter in the X. This is, evidently, a form of the competition idea of Aristotle, which had been previously elaborated by Weismann, and by Goldschmidt especially for sex determination. But Bridges based it on more direct experimental evidence and proceeded to use it as a working hypothesis to suggest further experimental approaches. His studies of the effects of different dosages of the small fourth chromosome on the phenotype of intersexes were designed to test the net effect of the genes in this chromosome on the sex of the individual. The results were inconclusive, but the method was used successfully by Dobzhansky and Schultz (1931, 1934) in studying the effects of various fragments of the X. When these fragments were added to intersexes (that is, 2X + 3A + fragment of X), they found that the "inert" (that is, heterochromatic) region was without effect, but that each of the very diverse euchromatic duplications shifted the degree of inter-sexuality toward femaleness. That is to say, not only does the X as a whole have a net female effect, but the numerous tested portions of it also have such an effect.

Similar tests with duplications for various autosomal segments have given no such clear-cut result, but evidence of another kind indicates that there are several autosomal genes that affect the sex of the individual. Several autosomal mutant genes change diploid females into male-like intersexes (Sturtevant, 1920, for *Drosophila simulans;* Lebedeff, 1934, and Newby, 1942, for *D. virilis;* L. V. Morgan, 1943; Sturtevant, 1945; Gowen, 1948, for *D. melanogaster).* Four of these genes are recessive, so that the supposition is that their wild-type alleles probably influence de-velopment in the *female* direction, rather than in the male direction as the whole set of autosomes does. They do, however, serve to indicate that there are several (many?) autosomal loci concerned.

In 1946 I found that intersexes are produced in hybrids between *Drosophila repleta* and *D. neorepleta,* The analysis indicated that neo-repleta carries a dominant autosomal gene which so conditions the eggs before meiosis that two repleta X's are insufficient to produce the full female phenotype and intersexes result—though two neorepleta X's, or one X from each species, produce normal fertile females. Here is evi-dence for an autosomal gene with the expected male effects; whether or not the unusual maternal effect is generally present in other Drosophila species remains to be determined. It may be noted that this particular gene would not be detected by the usual technique for testing the effects of duplications or deficiencies on triploid intersexes. This gene, since it operates before meiosis and fertilization, cannot be responsible for all the autosomal effects observed in the triploid experiments.

The earlier interpretation of the sex chromosomes of Drosophila, giving the male the formula "XO," meant that the male was haploid for this chromosome, and this was consistent with the facts of sex-linkage. When the work of Bridges and of Metz established that the normal male is XY, it became necessary to suppose that the Y lacked dominant alleles of the sex-linked genes. The study of the XO (exceptional) males by Bridges (1916) showed that, though phenotypically normal, they were sterile. It was shown by Stern (1929) that fertility is dependent on the presence of both arms of the Y, and also (Stern, 1926) that the Y nor-mally carries the wild-type allele of the sex-linked mutant bobbed, so it is not quite inactive genetically. Finally the cytological studies of Heitz and of Painter in 1933 showed that the Y is heterochromatic, like the "inert" right-hand portion of the X.

The work described up to this point has given a reasonably complete and consistent picture of the system of sex determination in Drosophila. It has, however, gradually become evident that this scheme cannot hold in detail for all other forms, even among those in which the normal system can be described as XX, XY.

It was long clear that the Y is not always necessary for fertility of the male, since it is missing in many groups of animals (including Pyrrhocoris, in which the X was first described, and some Diptera—in the same order as Drosophila). The first clearly inconsistent result came from the independent work of Westergaard and of Warmke and Blakeslee in 1939 on Melandrium. This plant is dioecious (has separate male and female individuals) and has clearly distinct X and Y chromosomes, with the female XX and the male XY. The study of induced polyploids and their offspring shows that Y is by far the most important element in the situation: individuals with a Y are male, those without a Y are female. The difference from the Drosophila system was unexpected, but most of us were inclined to minimize it as probably representing a special case. After all, most of the relatives of Melandrium are hermaphrodites, and in the genus itself each sex has clear rudiments of the organs of the other. An earlier indication that the system cannot be widely generalized came from the discoveries by Winge (1922, 1934) in the guppy, Lebistes. Here he showed first that the Y normally carries most of the genes that are responsible for the great variability in the color of the males. He then found that it was possible to produce strains in which the sex-determining mechanism was so modified that the female, rather than the male, was heterozygous for sex. This result was confirmed by Bellamy (1936) for another aquarium fish, Platypoecilus, where two species differ in this way. There is less clear evidence of similar instability of the system in Amphibia, and even in the Diptera there is evidence (Beermann, 1955, on Chironomus; Mainx, 1959, and Tokunaga, 1958, on Megaselia) that different pairs of chromosomes may function as sex-determiners (of the XX, XY type) in different races of the same species.

Finally, it has been found recently that mammals (man, Jacobs and Strong, Ford, *et al.*, 1959; mouse, Welshons and Russell, 1959) resemble Melandrium in that the Y is male determining. Other complications involved are beyond the scope of this book. There is now evidence (Ullerich, 1963) that the Melandrium system applies also to Phormia, one of the blowfly group, where XXY is male. It looks as though Drosophila is a rather exceptional type, even within the order Diptera to which it belongs.

As described in Chapter 6, sex-linkage of the type with the heterozygous female was known in moths and birds before discovery of the type with the heterozygous male, so it was clear from the start of the Drosophila work that there was a type that rested on a different sex-determining mechanism. It was with this type that Goldschmidt made the first attempt at an interpretation in terms of developmental genetics, based on his studies of *Lymantria dispar,* the gypsy moth.

He reported in 1912 on crosses between European and Japanese races of this moth, in which a Japanese female by a European male gave offspring in which both sexes were normal; the reciprocal cross, a European female by a Japanese male, gave normal F_1 males, but the F_1 females were more or less male-like—a condition for which Goldschmidt proposed the term *intersexuality.* After testing these hybrids in various combinations, Goldschmidt concluded that there are two opposing tendencies: F (for femaleness), inherited strictly maternally, and M (for maleness), inherited with the X. One X is insufficient to outweigh F, and FM is a normal XY female; two X's are enough to outweigh F, and FMM is a normal male. In the Japanese race, both F and M are strong; in the European race, both are weak. The intersexes then have a weak F and a single strong M, which is enough to shift the development partially, but not completely, in the male direction.

Goldschmidt's later work with this material was based on crosses with a great many geographical races, especially from Japan, where great diversity in "strengths" of F and M was found. These experiments were summarized in 1933. The original interpretation was confirmed, and greatly expanded and elaborated, in the later papers. It is not easy to evaluate the work. It has certainly been widely cited and acclaimed and has served to focus attention on problems having to do with genes and development. Nevertheless, some of us have serious doubts about it.

One doubt centers on the inheritance of F. This property goes strictly through the female line, without dilution or segregation. At one time Goldschmidt concluded that it was carried in the Y chromosome and exerted its influence on the eggs before meiosis, so that each egg was female in potentiality even if it lost the Y in the polar body. Later experiments seemed to Goldschmidt to disprove this, and he concluded that F is not chromosomal at all, but is cytoplasmically inherited. We are left with no explanation of the earlier experiments that were taken to show that F is in the Y. This contradiction remains unresolved.

The interpretation is in appearance a quantitative one and is often so described, but there are no quantitative data. The numerical values are arbitrarily assigned hypothetical ones: a value of "80" assigned to a single M does not refer to any measured or defined units. The papers contain numerous curves, representing specific hypotheses about the course of development, but these are also arbitrary and not based on any measurements.

The papers contain accounts of a large number of crosses, descriptions, and photographs of many intersexes. One cannot fail to be impressed by the extent of the work—but I confess that I should be more impressed if there had been use of more powerful genetic and cytological techniques, and more attempt to get objective quantitative data.

Another type of sex determination, the understanding of which began to emerge before the discovery of the sex chromosomes or even of chromosomes, occurs in the honeybee and other Hymenoptera and in a few other groups of animals.

Dzierzon, like Mendel, was a Silesian priest. A contemporary of Mendel, he suggested in 1845 that the males of the honeybee arise from unfertilized eggs, the queens and workers from fertilized ones. This view was at first opposed, but came to be generally accepted—especially when modified to state that the females are diploid, the males haploid.

The Hymenoptera are not easy cytological subjects, and the chromosome numbers were long in doubt. In fact, the first clear cytological demonstration of male haploidy and female diploidy was made by Schrader (1920) in the whitefly, Aleurodes (Homoptera). The spermatogenesis of the bee and hornet (Meves, 1904) showed that there is an abortive first meiotic division in which a small enucleate cell is budded off, with the result that each sperm receives an unreduced (haploid) complement of chromosomes.

There were numerous attempts, including one by Schrader and Sturtevant, to bring this system into line with that of Drosophila, but these were abandoned after the work of Whiting and his group on the parasitic Hymenopteran, Habrobracon. In 1925 Whiting and Whiting showed that, if the parents are related, diploid males may be produced in considerable numbers. By 1939 it was shown that there is a single series of multiple alleles (Bostian, 1939; Whiting, 1943)—at least nine members of the series are known—of such a nature that a heterozygote carrying any two different alleles is a female, while any individual with only one (either a haploid or a homozygous diploid) is a male.

This same type of sex determination has been established for the honeybee, but seems not to hold (at least in unmodified form) for some of the other Hymenoptera.

Still another type of sex determination is to be found in the mosses and liverworts, where the haploid generation is sexual. It was shown by the Marchals (1906, 1907) that, in certain mosses, regeneration from the diploid sporophyte (which reproduces by asexual spores) led to the production of sexual gametophytes. These were diploid, and were shown to be hermaphroditic, although the normal haploid gametophytes in these species are either male or female, as are homozygous diploids.

The conclusions suggested were confirmed by the cytological work of Allen (1917) on the liverwort Sphaerocarpos. He showed that the sporophyte has an unequal pair of chromosomes, of which the larger (called X) is present in the (haploid) female gametophytes, while the smaller (Y) is present in the male gametophytes.

Baltzer (1914) has shown that in the marine worm Bonellia sex is determined by the environment, and not by genetic means. This form has perhaps the most extreme sexual dimorphism known, the male being a minute degenerate creature that lives as a parasite in the female. Baltzer has shown that the larvae are not differentiated sexually. Those that settle on the sea bottom develop into females, while those that settle on the proboscis of a female develop into males.

There are many special devices in connection with several of the different types of sex determination, but even a simple catalogue of the more interesting ones would be out of proportion here.

It has sometimes been felt that the determination of sex offered the best opportunity for the study of the manner of action of genes, and the results described here have contributed largely to our understanding in this field; it now appears, however, that it will be more profitable to study simpler situations, and it is to these that attention is now more often directed (Chapter 16).

CHAPTER 14

POSITION EFFECT

In the absence of radiation, one mutation appeared in Drosophila with sufficient frequency to be worth detailed study: namely the case of Bar eye. Although the analysis was successful, the situation turned out to be too special to serve as a basis for any general picture of mutation, but it did lead to the discovery of the "position effect," which has played a large part in later developments.

Bar is a sex-linked dominant that reduces the size of the eye. It was studied intensively by Zeleny and his students, especially with respect to the effects of temperature on the size of the eye, as measured by counting the numbers of facets. In the course of these experiments it was noticed by May (1917) that Bar stocks occasionally revert to wild type. This phenomenon was studied by Zeleny (1919, 1920, 1921), who found that about 1 in 1600 offspring from a Bar stock carries a wild-type allele (B^+). He concluded that the event occurs in females, late in the development of their eggs. He also found that there is a more extreme type produced, a type that he called ultra-Bar, which I later gave the name double-Bar. He showed that double-Bar stocks also revert to wild type and also may give rise to Bar.

Zeleny's evidence indicated that these mutations occurred in females near the time of meiosis, and consequently Morgan and I were led to an investigation of whether the mutations had any relation to crossing over. Our result was clear: 6 reversions were obtained, and all were crossovers between marker genes (forked and fused) lying on opposite sides of the locus of Bar and less than 3 units from each other. That is to say, all 6 reversions were in a class (the crossovers) that included less than 3 percent of the population (Sturtevant and Morgan, 1923).

I then carried out more extensive tests (Sturtevant, 1925) that confirmed this result, not only for reversion of Bar but for the production of double-Bar and for other changes, such as the production of Bar from double-Bar/wild-type heterozygotes.

The interpretation developed was that of "unequal crossing over," according to which it occasionally happens that one chromosome breaks just to the left of Bar, the other one just to the right of it, yielding two crossovers—one carries no Bar (reversion to wild type) and the other carries two Bars (double-Bar, which is the reason for renaming it). This interpretation was later substantially confirmed and extended by study of the salivary gland chromosomes by Muller, Prokofieva-Belgovskaya, and Kossikov (1936), and by Bridges (1936). It appears from these studies that Bar itself is due to a "repeat" of the salivary section including the seven bands of section 16A. In double-Bar this section is present in triplicate. In a homozygous Bar female the pairing evidently sometimes occurs thus:

$$\frac{15 \cdot 16A \cdot 16A \cdot 16B}{15 \cdot 16A \cdot 16A \cdot 16B}$$

Crossing over within the apposed 16A sections then gives rise to $15 \cdot 16A \cdot 16B$ (wild type) and $15 \cdot 16A \cdot 16A \cdot 16A \cdot 16B$ (double Bar).

The two types, double-Bar/wild type and Bar/Bar, each have the 16A section represented four times, but facet counts (Sturtevant, 1925) showed that the former regularly has about 30 percent fewer facets than the latter. That is to say, three 16A sections in the same chromosome, and one in its mate, are more effective in reducing facet number than are two in each chromosome. This was at the time a wholly unexpected result, since all previous data had indicated that the position of a gene in the chromosome had no effect on its activity. The position effect here demonstrated has since been found to be rather widely distributed, and it is still being actively studied for its bearing on questions about gene action.

Later results (Dobzhansky, 1932; Bridges, 1936; Griffen, 1941; E. Sutton, 1943, and others) indicate that the original Bar phenotype is itself a position effect, due to the presence of a 16A section that is removed from its normal 15 neighbor, rather than directly to the dosage effects of the duplication. It may be surmised that the originally discovered position effect is due to a greater effect on the rightmost 16A section of double-Bar, since it is now still further removed from 15.

It has gradually become evident that there are two essentially different types of position effects, designated by Lewis (1950) as the S-type (stable) and the V-type (variegated). Bar represents the S-type, and this will accordingly be discussed first.

A position effect not dependent on a chromosome rearrangement was discovered by Lewis (1945). The two mutants Star (dominant) and asteroid (recessive) have very similar phenotypes and lie adjacent to each other. Lewis studied the heterozygote (Star/asteroid) and recovered from it both the wild type and the double mutant, Star asteroid. If we compare the two kinds of double heterozygotes—the *cis* type (Star asteroid/wild type) and the *trans* type (Star/asteroid)—it is clear that the *trans* type differs more decidedly from the wild-type phenotype than does the *cis* type. This case is complicated by the dominance of Star, but more recent examples have shown that the principle illustrated here is a general one in such cases: the *cis* heterozygote (carrying a normal unmutated chromosome) is more nearly wild type in phenotype than is the *trans* (Bar is an exception to this rule).

Situations like this were soon found in which the dominance complication was absent. The first of these was reported by Green and Green in 1949, for the mutant lozenge in Drosophila. Oliver (1940) had shown that females heterozygous for two independently arisen lozenge mutants, called glossy and spectacled (that is, glossy/spectacled), had the typical phenotype of the lozenge series but gave some wild-type chromosomes that were always crossovers for outside markers. Oliver failed to detect the contrary crossover and was therefore in doubt as to the significance of the result. Green and Green used glossy (lz^g) and two new independently arisen types (lz^{BS} and lz^{46}). They were able to recover from each of the double heterozygotes of the *trans* type, both the wild type and the double mutant; all these events were again associated with crossing over between outside markers. These results showed the sequence in the chromosome map to be lz^{BS} lz^{46}, lz^g. When the six possible heterozygotes for two mutants were made up, it was found that all three *cis* types were wild type in phenotype, whereas all three *trans* types (lz^{BS}/lz^{46}, lz^{BS}/lz^g, lz^{46}/lz^g) were lozenge in appearance.

The same kind of result was soon demonstrated for several other series of independently arisen alleles—for vermilion and for beadex by Green, for white and for bithorax by Lewis, and for several other series by other investigators. The case of white was particularly unexpected, for this had long been the type case of multiple allelism, and it became clear that the then-current hypothesis must be revised.

Multiple alleles had been supposed to represent changes in a single original gene, and there were two criteria for their recognition: they occupied the same locus in the chromosome and were not separable by crossing over; and their heterozygote *(trans* type) was mutant with respect to their common recessive phenotype, since neither carried the wild-type allele of the other. With the discoveries noted above, these two criteria were shown not to agree. In such cases, which came to be called "pseudoallelic" (Lewis), the *trans* heterozygote is mutant in phenotype (the mutants do not complement each other), but both the wild type and the double mutant can nevertheless be reconstituted by crossing over. Evidently, each mutant carries the wild-type composition that the other has lost, but the section of chromosome that includes them is a functional unit that must be intact in at least one chromosome to produce the wild-type phenotype.

This conclusion, which has been shown to apply to many (most ?) loci in many (all ?) organisms, has had a very wide influence. A minor consideration is that it has complicated the terminology of the subject in several ways. The symbolism for genes had grown up on the basis of the older view, and it is still not clear what will be the most effective compromise. The older terms *gene, allele,* and *locus* are now in a fluid state so far as current usage is concerned, and several newer terms are in general use: *cistron* (Benzer) to denote an area that must be intact (that is, in the *cis* form) to produce the wild-type phenotype, and *site* (or *recon* of Benzer) to denote the smallest unit separable by recombination. It is still not clear what will be the most convenient system of terminology; probably it will depend on developments in the study of the genetic coding system (Chapter 16).

There has also been much discussion of the implications of the position effect for the basic theory of genes and their effects on development. The most extreme view is that of Goldschmidt (1946), who suggested that the whole idea of genes be given up—the chromosome being a single developmental unit and all mutant effects being due to rearrangements (usually minute) of its parts, with resulting position effects. This view has few adherents, but at one time did figure largely in the literature of genetics.

In general, the frequency of recombination within a cistron is very low, and this is the reason why the phenomenon was overlooked in early work. It is still one of the limiting factors in the study of higher organisms. But in microorganisms, and especially in bacteriophage, it is possible to develop methods for studying recombinations that occur only with very low frequencies. It is largely for this reason that current studies in the field, leading to what is called "fine-structure analysis," are often carried out with such material. These studies (for example, see Benzer, 1961) are outside the scope of this book.

The second or "V-type" position effect probably is different in kind from the S-type. Most examples are associated with chromosome rearrangements induced by X rays. Muller reported in 1930 on certain "eversporting" types in which dominant genes were lost or inactivated in some cells of individuals carrying rearrangements, producing irregularly spotted patterns for eye color, body color, or other mutant types. It was evident that these spots were due to failure of action of genes near the break-points of the rearrangements. When more cases accumulated it became clear, as pointed out by Schultz (1936), that the inactivation usually occurs when genes in euchromatin are brought near heterochromatin or, less often, when genes in or near heterochromatin are brought nearer to euchromatin.

Dubinin and Sidorov (1935) described a translocation between the third and fourth chromosomes of Drosophila, with the breakpoint in III near the locus of hairy; this locus was brought near the heterochromatin of IV. They showed that the h^+ gene in this translocation exhibited reduced dominance over the h mutant allele. They were able to get crossing over between the translocation point and the locus of hairy and thus to recover the h^+ allele in a normal chromosome; its usual complete dominance was at once restored. A different h^+ allele was also introduced into the translocation chromosome by crossing over, and at once acquired the reduced dominance. In the same year, Panshin obtained similar results from another translocation that affected the dominance of the cu^+ gene.

These results furnished proof that these position effects did not depend on any transmissible changes in the h^+ or cu^+ genes but on an interference with their developmental effects.

There was one uncertainty about the phenomenon: Were the genes in question lost or only inactivated in the "mutant" areas? There is no increase in frequency of germinal losses, that is, the next generation from variegated individuals is variegated, not pure for the recessive alleles. Evidently this means either that the process does not occur in the germ line, or that it is reversed in the formation of the gametes. With the discovery of the same phenomenon associated with translocation in Oenothera (Catcheside, 1939, 1947, and later in mice by L. B. Russell and others) it became probable that the inactivation is reversed at meiosis, since there can scarcely be anything other than meiosis common to the history of the germ line in Drosophila and in Oenothera. That is to say, the V-type variegation is due to a suppression of the phenotypic effects of genes that are still reproducing in the usual manner at each cell division, so that reversal of the effect is always possible, though it usually does not occur except at meiosis. One may surmise that reversion is somehow associated with the uncoiling and lengthening of the chromosomes in the meiotic prophases.

Muller found in 1930 that this type of variegation might affect the action of several genes newly brought near heterochromatin. This phenomenon was studied by Gowen and Gay, by Patterson, and by Schultz;

the most detailed and illuminating studies were by Demerec (1940, 1941) and by Demerec and Slizynska (1937). These studies showed that there is a "spreading effect." If heterochromatin is represented by the symbol H and if a series of wild-type alleles A, B, C, and so forth be brought next to it in the sequence $HABC$, then the suppression of gene activity proceeds from H; A is inactivated first, then B, then C, and so on. Tissue with inactivation of A alone or of both A and B may occur but not with A active and B inactive. It appears that there is no skipping of genes, that is, there seem to be no genes immune to the effect. (A supposed exception to this rule reported for the fourth chromosome has been found to be based on an incorrect map of the fourth chromosome.)

The most likely interpretation is that there is a progressive inhibition of the production of gene products but not of gene replication; that is, in modern terms, RNA is not produced, but DNA replication does occur. One possible interpretation is that the timing of the DNA replication is retarded as it seems to be in heterochromatin; these matters are, however, beyond the scope of this book.

It is, in fact, premature to formulate any definitive scheme for the V-type position effects, since several facts remain to be further analyzed: the effect of removal from heterochromatin upon genes normally in or near it (possibly an inhibition of suppressors normally present ?);* the striking effects of temperature and of the number of Y chromosomes present (both reported in 1933 by Gowen and Gay); the occurrence of dominant V-type effects, and numerous other unexplained relations. These are now under active study in several laboratories, and there can be no doubt that the V-type effects will contribute largely to future ideas about the nature of gene action in development and differentiation.

---

* The most studied example of a gene normally located in or near heterochromatin that shows variegation when removed from most of this heterochromatin is that of cubitus interruptus (ci). It was shown by Dubinin and Sidorov (1934) that approximately half of the translocations that involve the fourth chromosome lead to a weakening of the dominance of ci^+ over the mutant ci. This case has been studied in great detail, especially by Dubinin and Stern and their co-workers. There are many interesting observations, some of which are rather puzzling, but it does not seem (to me, at least) that they have led to any close insight into the nature of such cases—in part because the nature of the ci phenotype makes it difficult to study.

CHAPTER 15

GENETICS AND IMMUNOLOGY

Both genetics and immunology deal with highly specific and very numerous reacting substances, and in both fields, efficient methods are available for analyzing the effects of these substances individually. It was, therefore, of considerable importance when the two approaches began to be applied to the same system. So far as genetics is concerned, this joint attack led to some of the first hopeful advances in the study of the manner of action of genes.

The beginning of this interaction dates from the discovery of the human blood groups by Landsteiner (1900). Early attempts at the clinical use of blood transfusion had frequently led to disastrous effects on the patient, so the practice had been given up, until after the work of Landsteiner and his followers was widely known and understood.

Landsteiner knew of the occurrence of reactions when blood of different species was mixed, and so he was led to look for differences between the bloods of individuals of the same species. He separated the red cells and the serum from the bloods of a number of human subjects and made cross-tests. He found that, in many combinations, such mixtures led to the agglutination of the red cells, and he recognized three different kinds of individuals. In 1901 he showed that there are two kinds of *agglutinins* (α and β) in the sera, and two kinds of *agglutinogens* (A and B) on the cells. When the fourth possible type of individual was found by Decastello and Sturli in 1902, the now familiar "ABO" blood-group system emerged. There were, however, several different systems of naming the groups, which were confusing at the time and still make some of the earlier literature difficult to follow. Table 2 shows the relations:

TABLE 2

Current Designation	Jansky System	Moss System	Agglutinogens on Cells	Agglutinins in Serum
O	I	IV	None	α, β
A	II	II	A	β
B	III	III	B	α
AB	IV	I	A, B	None

It was suggested by Ottenberg and Epstein (1908) that the blood groups are inherited, but the family data of these blood groups were not extensive enough to be convincing. However, von Dungern and Hirszfeld (1910) did succeed in demonstrating the point; they showed that any agglutinogen present in an individual was present in at least one of his parents (A and B are each dependent on a single dominant gene). They supposed that these two genes were independent in inheritance, as their data indicated. On this basis, Group O had the composition *aa bb*; Group A included both *AA bb* and *Aa bb*; Group B included *aa BB* and *aa Bb*; Group AB was of four kinds—*AA BB, Aa BB, AA Bb,* or *Aa Bb.*

That this view was incorrect finally became evident as an outgrowth of the observations of L. and H. Hirszfeld (1919). The Hirszfelds were army physicians in the Balkans during World War I and determined the blood groups of large numbers of soldiers of diverse races and nationalities in the armies there. They found clearly significant differences in the relative frequencies of the four groups among the sixteen races and nationalities studied, thus beginning the use of blood-group determinations in the study of population problems.

As the data from various populations accumulated, it was natural to compare them with the proportions expected from the algebra of random mating populations (see Chapter 17). I am certain that I was not the only one to realize that these proportions were not in agreement with the von Dungern-Hirszfeld scheme; but this did not seem surprising, since the algebraic analysis assumed random mating; and it is obvious that large human populations do not mate at random, for economic, religious, and geographical reasons.

The discrepancy led Bernstein (1925) to try other genetic interpretations, and he found that a system of triple alleles with only one locus concerned did give equilibrium frequencies in agreement with the observed frequencies, in populations with very different absolute frequencies. According to this scheme, the four groups have the following genotypes: O is *OO*, A is *AA* or *AO*, B is *BB* or *BO*, AB is always *AB*.

There was a serious difficulty here, as Bernstein recognized: On this scheme AB and O can never be related as parent and offspring, yet the published pedigrees included many examples of such relationships.

In 1929 Snyder presented a summary of the published records for the offspring of one of the critical matings—that between O and AB parents. If these are divided into two groups, those published before and those after Bernstein's paper, the totals are as shown in Table 3.

TABLE 3. OFFSPRING RECORDED FROM MATINGS BETWEEN O AND AB

Published Records	O	A	B	AB	Number of Papers
Von Dungern and Hirszfeld, 1910	2	2	2	3	1
All authors up to 1925 (including von Dungern and Hirszfeld)	27	80	59	24	18
All authors, 1927–1929	2	228	234	1	6

On the von Dungern-Hirszfeld interpretation, if nonrandom mating is assumed, no definite proportions can be calculated, although the second row certainly includes too few AB individuals; on the Bernstein interpretation the expected proportions are $0 : 1 : 1 : 0$. The three exceptions in the last row of the table are all listed as being under suspicion of illegitimacy. The 51 in the second row were recorded in eleven of the eighteen papers that are summarized.

This tabulation raises some disturbing questions. One has the uncomfortable feeling that observers see and report only what they expect to find. The most probable interpretation is that the methods of typing were improved. A study of the history of the clinical use of transfusion might be interesting, since the indicated frequency of misclassification before 1925 should have led to numerous unfavorable transfusion reactions.

The Bernstein interpretation has been consistently confirmed in more recent studies and is now fully established.

In one respect the ABO system is unusual in immunological studies, because the agglutinins are normally present in the blood of all individuals that lack the corresponding agglutinogens. The usual situation is that the effective serum components—*antibodies*, which may cause agglutination (as in the ABO system), hemolysis, or other reactions—are produced only as a result of the previous introduction of the corresponding *antigen* (agglutinogen in the ABO system) into the blood of an individual that does not itself produce it.

The extensive studies of blood antigens other than A and B in man, and those carried out on other animals, have most often utilized such *induced* antibodies.

If human blood cells are injected into a rabbit, a series of antibodies that will react with any human red cells will be produced in the rabbit's serum. Landsteiner and Levine (1927) carried out this experiment, using a series of rabbits, each injected with cells from one of a series of human donors. They then treated each antiserum with cells from several different donors separately. This procedure resulted in the absorption of all the general antibodies against all human red cells, but in some cases the treatment left antibodies that would still react with the cells from other individuals. They were able to show that there were two such reactive substances in the cells, which they called *M* and *N*. They found three types of individuals—*M*, *N,* and *MN*. These substances have been found to be dependent on a pair of allelic dominant genes: no "inactive" allele, corresponding to the O of the ABO system, is known.* This paper marks the beginning of the study of specific induced antibodies to human blood cells, which has been greatly extended since then.

A modification of this technique was used by Landsteiner and Wiener (1940). They injected red cells from a Rhesus monkey into guinea pigs, and carried out absorption with red cells from different human subjects. The resulting absorbed antisera were, in some cases, reactive with cells from other human subjects. They recognized two types of individuals, Rh positive and Rh negative, the symbol being derived from the name of the original donor species.

The clinical importance of the Rh antigens arises from the fact that human subjects who are Rh negative may develop Rh antibodies which lead to transfusion reactions, if the subjects have previously received transfusions from an Rh-positive donor (Wiener and Peters, 1940)—or, more often, if an Rh-negative mother has had an Rh-positive child (Levine and Stetson, 1939). In the latter case, it is evident that Rh antigens from the fetus enter the maternal circulation, and there induce the formation of anti-Rh in the serum. This antibody may, in turn, enter the

---

* This series of alleles has since been extended by the discovery of antibodies to "S," and to other properties. Many alleles are now known.

circulation of a later Rh-positive fetus. This results in the condition long known as "erythroblastosis fetalis."

Most of the later work in this field has been based on the study of antisera from such sensitized mothers. The work of Levine and Wiener and their co-workers in this country, of Mourant, Race, Sanger, and their co-workers in England, and of others elsewhere has led to a detailed and complicated analysis of the Rh system—now known to include a whole series of alleles. This work, which is of great clinical, anthropological, and genetic importance, is outside the scope of this book. It may be pointed out, however, that the literature is greatly complicated by the widespread use of two radically different systems of nomenclature, due to Wiener and to Fisher, respectively.

There are a series of other systems of red-cell antigens now known in man; these are also not further discussed in this book.

Absorption methods have been used in the study of the red-cell antigens in several other animals besides man. One of the first genetic studies of this kind was carried out by Todd (1930, 1931) with fowl. He injected the cells of various chickens into other chickens and pooled the resulting antisera. He then absorbed these polyvalent antisera with cells from one or more other birds and tested the resulting absorbed sera against the cells of still other individuals. Among three families of chicks, each from a single pair of a single strain of Plymouth Rocks, the results showed that no bird had antigens that were not present in one or the other of its parents, and that within each family (with 17, 18, and 13 chicks respectively) there were no two chicks of identical antigenic composition.

Here then, within a single strain of a single breed, was a great diversity of antigens; and each was dependent on a dominant gene, with no interaction in the production of the phenotype—either between alleles or between genes at separate loci. This remarkable result was soon interpreted to mean that the antigens were close to immediate gene products, and might furnish useful materials for the study of the action of genes, relatively free of the complications of developmental interactions. It is not clear who first formulated this idea; I first heard it in conversation with Haldane in the winter of 1932–1933. However, the results of this assumption have been of far-reaching importance in the study of the developmental effects of genes (Chapter 16).

The lack of interaction between the products of different genes in the determination of antigen specificities has been found to be a very general relation; there are, however, a few exceptions. Irwin and Cole (1936) reported two cases in species hybrids in doves and pigeons. They crossed

Ring Doves to Pearlnecks and to domestic pigeons, and in both cases found that antisera to F_1 cells were not completely exhausted for antibodies by successive absorption by cells from both parent species. In both cases these F_1 birds were backcrossed through several successive generations to one of the parent species, and several genetically distinct species-specific antigens of the usual noninteracting type were isolated. The "hybrid substance" was, in each series, related to particular ones of these, but as yet the nature of this relation is not entirely clear.

These, and a few more recently discovered examples, do show that interaction of gene products may occur; but the rule is still valid in the great majority of cases, even in the dove and pigeon hybrids. There are also a few clear exceptions in intraspecific crosses. This rule is, as formulated by von Dungern and Hirszfeld in 1910 for the human ABO groups, and by Todd in 1930 for his fowl, that no individual carries a red-cell antigen that was not present in at least one of his parents.

A related series of studies concerns the fate of grafted tissues in vertebrates. The first clear genetic result here was that of Little and Tyzzer (1916). They studied a tumor that could be successfully transplanted into any mouse of a strain of waltzing mice (in which the tumor had arisen spontaneously), but failed to grow in mice of an unrelated strain. When these two strains of mice were crossed it was found that the tumor would grow in the F_1, that is, susceptibility was dominant. But when F_2 mice were tested, only 3 of the 183 tested individuals were susceptible. They concluded that several (about 7?) independently segregating dominant genes must all be present in an individual to make it susceptible. This conclusion, checked in various ways, has since been established for many transplantable tumors—with cases on record for strains differing in one, two, or more genes necessary for the growth of particular tumors.

A similar situation has been studied, using normal tissues for transplantation. It has long been known that most normal mammalian tissues can be successfully transplanted to other parts of the same animal (autotransplants), but that transplants to other individuals are usually unsuccessful. There was evidence that the chance of success was better, though still poor, if the donor and host were closely related. The genetic analysis of this relation dates from the work of Little and Johnson (1921). They used inbred strains of mice, that were largely homozygous, and transplanted spleens. They found that such transplants were usually successful within an inbred line but not between separately inbred ones. When two such inbred lines were crossed, the parental strains would not accept transplants from the F_1, but the F_1 would accept those from either of the parental strains. These results were not very extensive nor wholly convincing, but they were fully confirmed on a large scale by Loeb and Wright (1927), using a series of long-inbred lines of guinea pigs.

This general approach has since been carried much farther in mice, especially by Snell, who has succeeded in genetically isolating a series of genes concerned with graft compatibility, and in locating these on the linkage maps.

One of the most striking results in this field was an outgrowth of the extensive work on the red-cell antigens of cattle. The existence of a great individual diversity in cattle antigens was shown by Todd and R. G. White in 1910, but the genetic analysis was begun by Ferguson (1941) and extended by a group including Irwin, Ferguson, Stormont, and Owen. In 1945, Owen reported on a pair of twins, one of which had a Guernsey sire, the other a Hereford sire. He showed that each of these twins had antigens that could only have come from the sire of the other and, similarly, had two kinds of red cells. That is to say, each had a kind of cell proper to its own genetic composition and another kind proper to that of its twin. This was interpreted as being due to the anastomosis of blood vessels in the placenta, already shown by Lillie (1916) and others to result in the passage of hormones from male embryos to their female co-twins. The new results indicated the reciprocal passage of cells ancestral to red cells; since the cells characteristic of a twin were found to persist into adult life, it followed that these foreign erythropoietic cells persisted and reproduced in the recipient animals.

The consequences of this finding were far reaching, and have led to developments in the study of the immunological basis of individual specificity, acquired tolerance to foreign tissue, and other topics (Owen, Medawar, and their co-workers).

CHAPTER 16

BIOCHEMICAL GENETICS

There have been two chief biochemical approaches to the study of genetics—through the biochemical study of the effects of gene substitutions, and through a direct attack on the chemical nature of the genetic material itself. Both approaches have been highly successful in recent years, but both went through a rather long period of slow development that was often rather discouraging.

The study of the biochemical effects of genes may be dated from the work of Garrod on alkaptonuria in man. In 1902, he concluded that this condition is an inherited one and that it is due to an alternative pathway in the metabolism of nitrogenous materials, leading to the excretion of homogentisic acid—rather than its further degradation product, urea—in the urine. He consulted with Bateson on the genetic question, and in 1902 the latter discussed the case, giving what appears to be the first suggestion of gene action in terms of "ferments" (enzymes) and of a recessive owing its properties to the absence of a particular "ferment." Later, Garrod (1908) discussed the case in much more detail and concluded that the condition is due to the blocking of a particular enzymatically controlled metabolic reaction, leading to an accumulation and excretion of the substrate (homogentisic acid) normally degraded by this reaction. He deduced the probable earlier compounds produced in this degradation and administered them to alkaptonuric patients. They were degraded to homogentisic acid, thus showing that the mechanism was interrupted at that one point but was still capable of functioning up to that point.

He also drew similar conclusions concerning some other less fully analyzed biochemical defects in man such as albinism, cystinuria, and porphyrinuria.

This work was early recognized as of great importance in the study of the chemistry of metabolism; it illustrated the principles of blocking of specific metabolic pathways, and the resultant accumulation of intermediate products, in intact living organisms. In 1909 the results were discussed by Bateson in what was the standard general work on genetics—required reading for every geneticist. He pointed out the general similarities between gene action and the action of enzymes, especially in their specificity and their effectiveness in very low concentrations. He concluded that genes often act through the production of enzymes but was doubtful that they could always do so, since he found it difficult to suppose that such a character as brachydactyly could be caused by the addition of an enzyme to the system. He also urged that the enzymes in question should be thought of as gene products, not as the genes themselves.

Bateson was influenced by the work at Cambridge—evidently stimulated by him—on the correlated genetics and biochemistry of the melanins in mammal coat colors (Durham, 1904) and of the anthocyanins in flower colors (Wheldale 1909). Cuénot had discussed the effects of genes influencing the coat colors in mice in terms of chromogens and enzymes in 1902 and 1903. These studies were continued by others (especially by Gortner and by Onslow with melanins and by Scott-Moncrieff with anthocyanins), and led to rather detailed descriptions of the effects of particular genes in biochemical terms, with enzymes regularly assigned major roles.

In 1917 Wright brought together the chemical and genetic data on the melanins in mammals, comparing all the forms on which evidence was available and arriving at a general scheme that rested on the assumption that the known genes produced their effects by conditioning the presence and the specificity of a few enzymes.

There was a widespread reluctance, however, to draw conclusions about direct gene action, since most geneticists were inclined to lay great emphasis on the complexity of development and to conclude that this approach had little chance of throwing light on the direct action of genes. The view that it may sometimes be possible to study fairly direct products of gene action grew out of the studies on the antigens of red blood cells, as described in Chapter 15. This idea soon bore fruit.

It was shown by Morgan and Bridges (1919), in their study of the gynandromorphs of Drosophila, that the sex-linked genes are usually "autonomous"; that is, each separate part of the body develops according to its own genetic composition, with no visible effects on the composition of the rest of the body. This type of analysis has also made use of the twin spots arising from somatic crossing over (Chapter 8). I showed in 1920 that the mutant vermilion is an exception to this rule. In gynandromorphs that are mosaics for vermilion (v) and wild-type (v^+) tissue, it sometimes happens that genetically v tissue develops the v^+ eye color, evidently through the influence of something it acquires from the v^+ tissue.

The next step in the analysis of this type of phenomenon was made by Caspari (1933), using the meal moth Ephestia. In this form there is a recessive mutant (a), that has pale eyes and testes. Caspari made reciprocal transplants of testes between larvae of the two forms (a and a^+), using a technique developed by Meisenheimer (1909) for other purposes with other moths. The results of Caspari's experiments were that: (1) a testes implanted in a^+ hosts developed darker color, and (2) a^+ testes implanted in a hosts not only developed full color but also induced a darkening of the testes and of the eyes of the recipient. Evidently here, again, there was a transfer of something from a^+ tissue to a tissue; in more recent terminology, a of Ephestia, like v of Drosophila, is a *reparable* mutant.

In 1935, Beadle and Ephrussi adapted the transplantation technique to Drosophila, by removing imaginal discs from one larva and injecting them into another larva. These implanted discs persisted and at metamorphosis developed into the parts they would have produced if they had been left in the original donor—though these parts usually lay free in the body cavity and did not replace the corresponding parts of the recipient larva. Through the use of this technique, Ephrussi and Beadle were able to show that the eye discs of v and v^+ Drosophila behave like the testes of a and a^+ Ephestia; v discs implanted in v^+ hosts develop the v^+ color, and the eye color of v hosts is modified toward v^+ by the presence of v^+ implants.

A similar study using many other mutant types showed that most of them were autonomous in development, thus confirming the conclusions derived from gynandromorphs. But the recessive eye color cinnabar (cn) proved to be reparable, behaving quite like vermilion (cn^+ and the a^+ of Ephestia were later shown to be mutually substitutable). The two mutants, v and cn, are alike in phenotype, both lacking one of the two pigments (the brown one) that are present in wild-type eyes. When reciprocal transplants were made between vermilion and cinnabar larvae, it was found that vermilion discs in cinnabar hosts developed wild-type eyes, but cinnabar discs in vermilion hosts failed to develop the brown pigment. That is, cn hosts produce the v^+ substance, but v hosts do not produce the cn^+ substance.

Beadle and Ephrussi (1936) concluded that the two substances are sequential in the normal metabolic chain, and that the reactions may be represented thus:

$$\text{substrate} \rightarrow v^+ \rightarrow cn^+ \text{ substance}$$

This type of argument has since been applied to many systems and has proved to be one of the most powerful methods in the analysis, through the study of mutant types, of metabolic syntheses in intact organisms.

The study of this system was actively carried on by Ephrussi and co-workers (Khouvine and others), by Beadle and his co-workers (Tatum and others), and by others in several laboratories. These studies, culminating in the identification of the v^+ substance as kynurenine (Butenandt, Weidel, and Becker, 1940) have been described many times—for example by Ephrussi (1942).

The use of Drosophila for this kind of work was stimulated by the existence of a large number of mutant types, whose interaction, it was hoped, could be useful in future studies; but with the discovery that the great majority of these are autonomous it became evident that the transplantation technique was not useful in their analysis.

This realization led Beadle and Tatum to look for a more favorable object. They chose the fungus Neurospora. The techniques for handling this material had been worked out by Dodge and by Lindegren, who had also used it for interesting genetic studies. It had also been shown by Robbins that Neurospora can be grown on a relatively simple synthetic "minimal medium," containing a carbon source, small amounts of inorganic salts, and a single complex organic compound (biotin). On this medium the plant manufactures for itself the other organic substances that it needs—amino acids, vitamins, and other, still more complex, substances such as proteins. Beadle and Tatum reasoned that it should be possible to detect and to study any mutants affecting the ability to synthesize such substances, providing they were reparable. That is to say, if the substance not produced by a mutant could be supplied in the medium and then utilized, growth should occur, and genetic and biochemical studies could be carried out.

The plant has other advantages for this purpose. It is haploid, so that dominance and recessiveness do not complicate the analysis; it usually reproduces asexually by the production of numerous "conidiospores," which make it possible to work with very large numbers of individuals of identical genetic composition and also simplify the detection of large numbers of mutant individuals; sexual reproduction can occur, making possible a detailed genetic analysis of the mutants produced.

In their first paper on the subject, Beadle and Tatum (1941) reported the recovery of three mutant strains after X-ray treatment. These grew normally on a "complete medium" made by adding malt extract and yeast extract to the "minimal medium" described earlier but were (unlike the wild type from which they came) unable to grow on the minimal medium itself. Further analysis showed that they could be grown on minimal medium supplemented with pyridoxine (vitamin B_6), thiamine (vitamin B_1), or para-aminobenzoic acid, respectively. The type that required pyridoxine was shown to be inherited as a single-gene mutant type—as were the other two and a whole series of additional mutants that were soon reported, in papers quickly following the original account. The techniques were gradually improved in several ways and were developed so that spontaneous mutants (occurring with a very low frequency) could be studied, as well as those induced by X rays or ultraviolet light.

These studies were rapidly extended at Stanford (Beadle, Tatum, Mitchell, Horowitz, Houlahan, D. Bonner, and others) and other laboratories. They led to work with similar methods on other microorganisms, following the discoveries of sexual reproduction or similar phenomena leading to recombination in yeast (Winge, 1935); bacteria (Avery, MacLeod, and McCarty, 1944; Tatum and Lederberg, 1947); and bacteriophage (Hershey, 1946).

These studies have led to great advances in our knowledge of metabolic pathways and in our understanding of the mechanisms of gene recombination and of the ways in which genes act. These studies are outside the scope of this book, but mention must be made of the "one gene–one enzyme" hypothesis, which is now dominant in the study of gene action.

The biochemical approach to the study of the nature of the genetic material goes back to the work of Miescher (1871). He studied the nuclei of pus cells (and, later, of fish sperm cells) and described a substance that he called nuclein. The paper was submitted in 1869, but the editor, Hoppe-Seyler, held it for two years while he repeated some of the observations that had seemed to him to be rather improbable. There followed a series of papers on the subject by Miescher and others. In 1889, Altmann showed that the substance could be split into protein and nucleic acid—the latter being unusual among complex organic compounds in lacking sulfur, but containing much phosphorus. By 1895, Wilson could write:

> Now, chromatin is known to be closely similar to, if not identical with, a substance known as nuclein—which analysis shows to be a tolerably definite compound composed of nucleic acid (a complex organic acid rich in phosphorus) and albumin [protein]. And thus we reach the remarkable conclusion that inheritance may, perhaps, be effected by the physical transmission of a particular chemical compound from parent to offspring.

In the next year he suggested that it is the nucleic acid component that is responsible for heredity.

The existence of two kinds of nucleic acid was shown by Kossel in his studies of material derived from thymus and from yeast cells. These two types, long known as "thymus nucleic acid" and "yeast nucleic acid," are now known as deoxyribose nucleic acid (DNA) and ribose nucleic acid (RNA), respectively. Ascoli (1900) and Levene (1903) showed that both contain adenine, cytosine, and guanine, while the thymine of DNA is replaced by uracil in RNA. Levene also established the presence of deoxyribose and ribose, respectively, as the sugars present.

The early analyses of nucleic acids suggested that the four bases were, in each case, present in equimolar amounts, and the "tetranucleotide" interpretation was developed. According to this view, an essential component of the nucleic acid molecule is a unit composed of one of each of the four bases. Chargaff and his co-workers showed (1950 and later) that in DNA the bases need not be present in equimolar amounts, but that the amount of cytosine does equal that of guanidine, and that of adenine is equal to that of thymine.

With the development of a stain specific for DNA (Feulgen and Rossenbeck, 1924) and, later, of enzymatic methods for distinguishing DNA and RNA, it became possible to study the distribution of both substances in the cell. Feulgen (1937) showed that, in most cells, almost all the DNA is in the nucleus.

Another approach grew out of the studies of Griffith (1928) on the bacterium Pneumococcus. It had been found previously that this organism exists in a number of serologically distinct types (Neufeld and Haendel, 1910; Dochez and Gillespie, 1913, and others, later). The type specificity is due to differences in the mucopolysaccharides in the capsular envelopes of the bacteria, and in culture of the strains *in vitro* it may be lost, leading to "rough" strains that have lost both their type specificity and their pathogenicity. Griffith (1928) used a rough strain that had been derived from a Type II strain, which he injected into mice along with killed individuals of Type III. The mice died, and from their bodies Griffith recovered virulent Type III strains of Pneumococcus. This was the first example of transformation, as the phenomenon came to be called. Dawson and Sia (1931) succeeded in getting it *in vitro* instead of making the mixture in a mouse. The studies were continued, in an effort to isolate the "transforming principle" from the killed virulent strain, and in 1944 it was shown by Avery, MacLeod, and McCarty that this agent is in fact DNA—though the specificity conferred by it and inherited by the descendants of the transformed cells is due to a polysaccharide.

It had been generally supposed that the degree of specificity evidently present in hereditary material could only reside in proteins. Partly because of the tetranucleotide interpretation, it had seemed that the nucleic acids were too simple for this degree of specific diversity. With the breakdown of the tetranucleotide theory, and even more because of the Pneumococcus result, attention turned to DNA. The spectacular results in this field in recent years are not within the scope of this book, since they have been described many times, and are still being very actively developed. I can only mention the solution of the structure of DNA by Watson and Crick (1953) and the work on the RNA code by Nirenberg, Ochoa, and many others.

CHAPTER 17

POPULATION GENETICS AND EVOLUTION

The mechanism of heredity and variation is basic to the study of evolution and was therefore a major concern of Darwin and many of his followers, including Galton and Weismann. It was an interest in evolution that led several of the earlier Mendelians, such as de Vries and Bateson, to the study of heredity. But, with the discovery of Mendel's work in 1900, the development of the new methods caused a temporary lack of interest in the evolutionary implications. As Bateson put it in 1909:

> It is as directly contributing to the advancement of pure physiological science that genetics can present the strongest claim. We have an eye always on the evolution-problem. We know that the facts we are collecting will help in its solution; but for a period we shall perhaps do well to direct our search more especially to the immediate problems of genetic physiology . . . willing to postpone the application of the results to wider problems as a task more suited to a maturer age.

Evolution is concerned with changes in populations, rather than in individuals, and what was needed was an analysis of the effects of the Mendelian scheme on populations of interbreeding individuals. The beginning of this analysis is a paper by Yule (1902) in which he pointed out that, if the members of an F_2 population, segregating for a single pair of genes (A and a), interbreed at random, the three types of individuals (AA, Aa, aa) will be represented in the same proportions in the following generations. He also raised the question: What will happen if all the aa individuals are removed? His analysis here was in error, but it was corrected by Castle in 1903. Castle pointed out also that, if such selection ceases in any generation, the newly established proportions will then be stable.

Here was the essence of the basic formula of population genetics, though it was derived by a longhand method and was not stated in simple algebraic form. The result, without selection, was also derived by Pearson (1904), that is, for the case where $p = q$, where p = the frequency of A genes, q = the frequency of a, and $p + q = 1$.

The generalization that the stable frequency of genotypes is $p^2AA : 2pqAa : q^2aa$ was made by Hardy and by Weinberg, independently, in 1908. Both knew of Pearson's result. To Hardy, who was a mathematician, the generalized result seemed so self-evident that he commented: ". . . I should have expected the very simple point which I wish to make to have been familiar to biologists." That it was not familiar is shown by the fact that it had been seriously suggested that dominant genes would automatically increase in frequency in mixed populations.

The Hardy-Weinberg formula is strictly valid only if several conditions are fulfilled:

1. The population must be large enough so that sampling errors can be disregarded. As Hardy pointed out, any chance deviations in the values of p and q will be as "stable" in succeeding generations as were those of the preceding one.

2. There must be no mutation, since change of A to a or of a to A will alter the values of p and q.

3. There must be no selective mating.

4. There must be no selection, that is, A and a must have no differential effect on the reproductive capacity of individuals bearing them.

(In the requirements 2 and 4, the wording given is not strictly correct, since it is possible that balanced mutation and selection rates may exist—in which case there will be no net changes in the frequencies of A and a.)

These are rather stringent requirements, and it may be doubted if they are ever fully met; nevertheless, they are often approximated closely enough to make the formula useful in analysis of populations.

The further developments in this field have depended on the algebraic analysis of the effects of deviations from the four requirements. This development dates from Haldane's (1924 and later) analysis of the effects of selection. He determined the number of generations required to alter gene frequencies, as related to the intensity of selection. This was worked out for dominant and for recessive genes, both in haploid and in diploid organisms, and for sex-linked as well as for autosomal genes.

This analysis has been followed by detailed studies on the algebraic consequences of variations in each of the four requirements by Haldane, Fisher, Wright, and others. Perhaps the most useful general summary is that of Li (1955).

One of the early developments was a theory of the origin of dominance (Fisher, 1928). Fisher suggested that mutant genes are inherently neither strictly dominant nor strictly recessive but produce more or less intermediate heterozygotes. There are, however, numerous modifying genes that affect their dominance. Since most mutant genes have unfavorable effects on their bearers, an individual heterozygous for a mutant gene will leave more offspring if the modifiers it carries happen to make the gene more nearly recessive. This effect will be slight, since an unfavorable semidominant mutant will not persist long in the population. However, since mutants continually recur with a low frequency, the effects will be cumulative over very long periods. This hypothesis has played a large part in discussions of population genetics but has been criticized by Wright and others on several grounds. The postulated effect of modifiers is a second-order effect, and it seems likely that their frequencies in the population will usually be determined by more direct effects. Also, an alternative interpretation is that a favorable gene will undergo selection—probably largely among alleles of the gene itself—for a "factor of safety," so that it will be capable of producing an excess of its useful product in times of stress; and such an excess may be supposed to result directly in dominance.

The first attempt to assemble a coherent general account of the algebraic analysis of Mendelian population behavior was Fisher's book in 1930. Widely read and discussed, it certainly strongly influenced further developments. One of the elements it minimized was that of the effects of population size. This was analyzed by Wright in several papers, first summarized in 1932.

Wright pointed out that in small populations there is a possibility of chance shifts in gene frequencies which are not controlled by selection and that this may lead to the production of combinations of genes— sometimes favorable—that would have almost no chance of being produced in large populations. He suggested that the most favorable condition for rapid evolution is that of a large population that is split (geographically or otherwise) into a series of relatively small subpopulations, with gene flow between these possible, but strongly restricted. Under these conditions, what has come to be called "random drift" may produce more favorable combinations of genes in some subpopulations, and these will gradually spread to neighboring populations. More often, probably, the result of random drift will be a less favorable combination of genes; when this happens, the subpopulation will diminish and probably be replaced by migrants from neighboring areas. Selection still remains the determining element, but emphasis is placed on selection among subpopulations rather than solely among individuals.

These analyses are largely theoretical, based on laboratory experiments—not on the actual properties of naturally occurring wild populations. It has turned out, not unexpectedly, that such populations are difficult to study. The situations encountered are so complex that it becomes difficult to evaluate separate variables. Further, when a quantitative analysis is made, it is specific for the population studied and cannot safely be applied to other populations, even of the same species. In other words, generalizations are difficult and dangerous. One result often obtained is that new kinds of problems have been suggested for further exact algebraic analyses.

In spite of these difficulties, much progress has been made. The first attempt to coordinate the existing quantitative data on natural populations and interpret them in terms of algebraic studies, was the book by Dobzhansky (1937).

It had, of course, long been evident that inherited diversities do occur in natural populations. Such examples as separate sexes, or heterostyly in plants, are obviously special cases. But the occurrence of "sports" and of inherited slight differences in such things as size, cold-resistance and color, among other things, was also familiar. It was not so clear that there is a store of recessive genes suitable for exact study by Mendelian methods. Examples of these had also been found by many observers, but the first attempts at a quantitative determination of their frequency in wild populations seem to have been made by a series of Russian investigators, using Drosophila. This work was initiated by Chetverikov (sometimes transliterated as Tschetwerikoff) in 1926 and culminated in the work of Dubinin and collaborators (1934, 1936), who studied a large series of wild populations of *Drosophila melanogaster* collected in the Caucasus. They found that up to 16 percent of the second chromosomes in these populations contained recessive lethals. It had not been supposed that wild populations contained such high frequencies of unfavorable recessives, but further work by Timoféeff-Ressovsky, Sturtevant, Dobzhansky and co-workers, and others has confirmed the result for several species and also has, in some cases, shown even higher frequencies of lethals and other unfavorable recessives.

Out of this work grew the realization that natural populations are full of hidden genetic variability, most of it potentially unfavorable. The study of this "genetic load" is now being actively prosecuted, in part because of its importance in the practical breeding of animals and plants, and because of the increase in frequency of mutations in human populations that must be supposed to have followed increases in irradiation caused by medical and dental uses of X-rays and by exposure to radioactive fallout from atomic bombs.

Another approach to the study of the relation between genetics and evolution is through the use of species hybrids. This, as pointed out in Chapter 1, has a long history; but only with the development of Mendelian and cytological methods did it begin to yield really helpful results.

It was soon apparent that different species usually differ in many pairs of genes and accordingly give numerous recombination types in F_2; Mendel himself indicated this in his discussion of Gärtner's work. With the development of the multiple-gene interpretation of quantitative inheritance (Chapter 9), it became possible to give a more precise interpretation of the results.

One of the awkward circumstances about discussions in this field is the ambiguity in the use of the word *species*. There is no generally accepted definition, but most discussions have centered on the extent to which cross-sterility between populations is to be taken into account in deciding whether or not two groups are to be considered as separate species. The nature and origin of interspecific sterility has been recognized since Darwin as one of the major problems of evolution.

The genetic study of the nature of interspecific sterility is difficult. If sterility is complete, this very fact makes it impossible to study by conventional genetic methods; if it is partial, there usually arises a possibility (or certainty in some cases) that it involves so great a distortion of segregation and of the relative viabilities of the various recombination products as to render analysis difficult or impossible. When these complications are absent or unimportant in any given case it becomes questionable whether the data obtained are relevant to the general problem.

The outcome of crosses between distinct species varies widely from one case to the next. The eggs and sperm may never come together (through lack of mating or failure of pollen tube growth); they may fail to fuse even if they do come together; if fertilization is effected, some or all of the chromosomes derived from the sperm may be eliminated at cleavage in the foreign cytoplasm (Baltzer, 1909); cleavage and mitosis may be normal, with the development of the gonads of the hybrid blocked at almost any point—most often just before meiosis; meiosis may be abnormal or essentially normal; the gametes may be normal in structure and be functional but may give rise to some or many abnormal or sterile individuals in the next generation. In other words, there are many different kinds of mechanisms that prevent or hinder interspecific exchange of genes. The problem is, how do these arise?

There is probably no single general answer to this question, but there is an answer applicable to many cases, namely, polyploidy.

The study of polyploidy may be dated from the work of Boveri and others in the 1880's and 1890's on the two races of *Ascaris megalocephala*—univalens, with one pair of chromosomes in the germ line, and bivalens, with two pairs. There is some question about this being a simple case of polyploidy. The first unambiguous cases seem to be those reported by the Marchals (1906, 1907) in mosses (described in Chapter 13) and since studied in great detail by F. von Wettstein, and in Oenothera by Lutz, also in 1907 (see Chapter 10). Other more or less probable examples followed (Strasburger, Tischler, and others) together with cytological studies of meiosis in the triploids produced by crosses between diploids and tetraploids. The results were confusing and contradictory, until the analysis by Winge (1917), which began to clarify the situation.

Winge made a detailed study of the available data on chromosome numbers in plants and found that in many groups there was a basic number, with various multiples of this number represented in different species. He pointed out that, in a hybrid between two species, it sometimes happens that some (or all) chromosomes are sufficiently different so that they do not pair at meiosis, so that the resulting gametes do not all contain a single complete set of chromosomes, and partial, or essentially complete, sterility results. Now, if the chromosomes of such a hybrid are doubled (by chromosome division without an accompanying cell division), each chromosome will now have an exact mate, and meiosis can be expected to be normal—with a restoration of fertility, as had already been shown by Federley (1913). This process was demonstrated by Clausen and Goodspeed (1925) and Clausen (1928), in Nicotiana. *N. tabacum* (24 pairs of chromosomes) was crossed to *N. glutinosa* (12 pairs). The hybrid, with 36 chromosomes, was sterile and at meiosis showed only a few paired chromosomes. One hybrid individual was fertile, however, and was found to be a tetraploid, with 72 chromosomes that formed 36 bivalents at meiosis; this plant was fully fertile and bred true to type.

The terminology which was suggested by Kihara and Ono (1926) has been generally accepted and has helped to clarify the relations. They suggested that:

> Under polyploidy we must distinguish two phenomena, namely, autopolyploidy and allopolyploidy. Autopolyploidy signifies the doubling of the same chromosome set; allopolyploidy the multiplication of different chromosome sets brought together by hybridization.

This distinction has proved to be convenient and useful—though it is not of an all-or-none type, since intermediate conditions occur. Allopolyploidy has been found to be widespread in the higher plants, but is rare in animals. Muller (1925) suggested that this is because it leads to difficulties in sex determination in species with separate sexes. It now seems more probable that the difficulty lies usually with the crossing of new tetraploids to diploids and the production of relatively sterile triploids; self-fertilizing hermaphrodites can avoid this difficulty.

The relative (or complete) sterility of the triploids is, however, of importance in that it leads to an immediate effective sexual isolation of an allopolyploid from both its parental forms. In fact, this sterility often operates to make difficult the production of any backcross offspring at all (Karpechenko and others), presumably because of an interaction between the tissues of the style and of the pollen tube, since it is known that, in some autopolyploid series, haploid pollen functions best in diploid styles, and diploid pollen in tetraploid styles.

Polyploidy, then, does form one method of bringing about interspecific sterility, but even in the higher plants where it is relatively frequent, it must be considered a rather unusual cause that has little bearing on the general question.

It should be added that the study of polyploidy has been of importance in other directions in genetics. Its use by Bridges and others in the analysis of sex determination has been discussed in Chapter 13, and it has also contributed largely to our understanding of chromosome mechanics, as developed by many authors. There have also been extensive applications in the breeding of cultivated plants. One of the important events in this field was the discovery (Blakeslee and A. G. Avery, 1937) that doubling of the chromosome set in plants may be induced by the use of colchicine.

Winge's interpretation of polyploidy grew out of a comparison of the chromosomes of related species. Another type of conclusion was based on such comparative studies by Metz (1914 and later) with species of Drosophila and by Robertson (1916) with a series of genera and species of grasshoppers. In both of these groups there are rod-shaped chromosomes, and also (in other species) **V**-shaped ones. Both authors found that, if each **V** was counted as two rods, the haploid number of elements was constant within a given group: 5 in Drosophila, if the small dot (often difficult to see) is neglected; 7 in the grouse-locusts, and 12 in ordinary grasshoppers. The conclusion was drawn that, in general, the elements maintain their individuality (to a large extent at least) within such groups, and are separated, or united, in various ways in different species.

This conclusion was questioned by R. C. Lancefield and Metz (1921) as a result of studies on *Drosophila willistoni*. This species, like *D. melanogaster*, has a pair of rods and two pairs of **V**'s, but in melanogaster the rod is the X chromosome, whereas in willistoni, Lancefield and Metz showed that one of the **V**'s is the X.

This anomalous result was explained later as a result of studies on the mutant genes found in various species of Drosophila. Such studies, by Metz, Sturtevant, D. E. Lancefield, Weinstein, Chino, Moriwaki, and others, showed that mutants with phenotypes closely resembling those of melanogaster could often be found in other species, and in the case of simulans, which can be crossed to melanogaster (although the hybrids are all completely sterile), it was possible to show that many of these resemblances are in fact due to mutations of the same wild-type genes (Sturtevant, 1921). This conclusion rests on more indirect evidence for the other species, but, as more examples accumulated, one rule appeared: mutants in other species that closely resemble sex-linked mutants in melanogaster are also sex-linked—though the reverse relation does not hold so consistently. In retrospect, it is obvious that most of the clear exceptions to the reverse rule concerned mutants in *D. willistoni* and *D. pseudoobscura* that were sex-linked in those species but resembled autosomal mutants in melanogaster; in both of these species the X was known to be a **V**.

The obvious conclusion was implied by D. E. Lancefield (1922) but was not elaborated or made more specific until Crew and Lamy (1935) carried out a detailed comparison of the known mutants of pseudoobscura with the apparently similar types of melanogaster. This paper was not very critical and used terminology that made it difficult to understand, but the conclusions were fully confirmed and extended by the more substantial accounts of Donald (1936) and of Sturtevant and Tan

(1937). These results confirmed the conclusion of Metz and Robertson: the six elements of pseudoobscura do in fact correspond to those of melanogaster. The X of pseudoobscura is **V**-shaped, and one arm contains the material of the melanogaster X, the other that of the left limb of the melanogaster III. The remaining four elements of melanogaster are all intact but are separate. That is, both **V**'s of melanogaster have their two arms separated, and one of these (III L) is now the right arm of the X.

This type of comparison has been extended to several other species of Drosophila that have haploid chromosome numbers from 3 to 6 (summary and analysis by Sturtevant and Novitski, 1941). It appears that the 6 elements have retained their composition, with relatively few exceptions, but that within each element the sequences of loci are little more alike than would result from chance alone. In other words, inversions within an element have been frequent; translocations, or inversions including centromeres, have been rare, in the evolution of the genus, as they are within existing species. There are a few examples of persistent associations of closely linked genes, which may be due to chance or to the existence of favorable position effects, but such persistent sections are not common.

The studies just discussed lead to the conclusion that there is a long-time stability in the genetic basis of particular characters, but such a stability has often been questioned. Perhaps the most extreme formulation of this point of view is that by Harland (1936):

> The genes, as a manifestation of which the character develops, must be continually changing ... we are able to see how organs such as the eye, which are common to all vertebrate animals, preserve their essential similarity in structure or function, though the genes responsible for the organ must have become wholly altered during the evolutionary process, since there is now no reason to suppose that homologous organs have anything genetically in common.

This conclusion was based on solid facts derived from extensive species crosses in the genus Gossypium (cotton). I have elsewhere (Sturtevant, 1948) given my reasons for an alternative interpretation of these facts, based on the polyploid nature of cotton. The more recent comparative biochemical data also favor the idea of the great stability of genetic systems, since they show essential identity of some of the gene-controlled basic biochemical pathways in bacteria, fungi, and vertebrates.

It is true, however, that in many, probably most, loci there exist series of isoalleles (Stern and Schaeffer, 1943), which carry on the function characteristic of the locus, but with different efficiencies and different temperature characteristics and reactions to the presence of gene differences in other loci. These often lead to no phenotypic differences under normal conditions, and can only be studied by special methods. It seems probable that the efficiencies of those concerned with any given developmental system need to be properly adjusted among themselves to give a harmonious system, as postulated by Goldschmidt for "strong" and "weak" races of Lymantria (Chapter 13). But other equally effective systems are also possible, and may come to exist in related species. Disharmonies will then arise on species crossing, and there are examples that suggest that this is often the case (Sturtevant, 1948).

CHAPTER 18

PROTOZOA

Weismann's germ-plasm theory was based on the differentiation of germinal and somatic cell lines in many-celled organisms. In unicellular forms there is evidently no such cellular differentiation, and Weismann surmised that these organisms are potentially immortal—a view already expressed by Ehrenberg in 1838. This conclusion was questioned as a result of the growing knowledge of the occurrence and nature of conjugation—first noted in the ciliate Infusoria.

Conjugation was observed by Leeuwenhoek (1695) and by O. F. Müller (1786) but was usually misinterpreted until the work of Bütschli (1873 and later) and Engelmann (1876). The ciliates have two types of nuclei—macronucleus and micronucleus—which were earlier interpreted as ovary and testis, respectively. Bütschli and Engelmann made out their nature and found that the macronucleus was lost at the time of conjugation. They developed the view that these animals are not potentially immortal but have a definite life cycle that requires conjugation to renew the vitality of the line (which becomes senescent and dies out in the absence of conjugation). This speculation was based on observation of the structural changes.

The cytological details of conjugation were worked out independently by Maupas and by R. Hertwig in 1889. They described the degeneration of the macronucleus, the exchange of micronuclei with the fusion of those from the two individuals, and the reconstitution of the new macronucleus from a division product of the biparentally derived new micronucleus. With these results, it became possible to relate the process of conjugation to the better-known sexual reproduction of higher forms.

Maupas had already (1888) published his elaborate studies on the relation of conjugation to senescence and rejuvenescence and had furnished an extensive experimental basis for the ideas suggested by Bütschli and Engelmann. There followed a long series of studies (by Hertwig, Calkins, Woodruff, Jennings, and others) on this general subject (see summary by Jennings, 1929).

Jennings understood that sexual reproduction should lead to segregation, recombination, and increased variability; and he set out to study these. The material was difficult, especially because conjugation was erratic and uncontrollable in its incidence. This difficulty was overcome by the discovery of mating types in Paramecium (Sonneborn, 1937), leading to the study of the ciliates by Mendelian methods.

It was shown by Blakeslee (1904 and later) that morphologically similar strains of the fungus Mucor fall into two types (called "plus" and "minus"). Sexual reproduction occurs only between the two types, never between individuals of the same type. The similarity to those forms having distinct male and female gametes was obvious, and such examples (soon found in many lower organisms) were often discussed as extreme examples of sexual differences in which the structural differences of eggs and sperm were absent.

Sonneborn found that similar relations occur in *Paramecium aurelia,* the different asexually produced lines falling into plus and minus series, between which, under appropriate conditions, conjugation occurs quickly and often in a striking *en masse* way, when plus and minus strains are mixed. But the analogy with sex is somewhat strained by the fact that each member of a pair contributes a micronucleus to the other, so that the natural comparison is that of this migrant nucleus with the sperm of higher organisms; in other words, Paramecium may be thought of as hermaphroditic, yet mating types occur. The comparison to sexual types was still further strained when Jennings (1938 and later) found that, in *P. bursaria,* there are eight mating types, of such a nature that each will conjugate with any one of the other seven. Eight sexes?

The comparison of mating types to sexual differences had already been complicated by the earlier work on "relative sexuality" in algae. This development begins with the work of Hartmann (1925) on the brown alga Ectocarpus. Here the gametes are structurally different, one type (comparable to sperm) being an active swimming body, and the other (comparable to an egg) nonmotile. These gametes are produced only when two compatible plants are in contact. Hartmann found that each plant gave consistent results when it was repeatedly tested against another one; but it was not possible to recognize male and female plants. Instead, there seemed to be a graded series, from extreme male to extreme female. Plants between these extremes could function as females when paired with extreme males, and as males when paired with extreme females.

Similar systems were found in other algae and were studied in detail by Hartmann and his students. One of the most striking cases, that of Chlamydomonas, was studied by Moewus, who developed an elaborate biochemical and genetic interpretation. This work at one time figured largely in the literature of genetics, though there were doubters almost from the beginning. It has now been very generally concluded that the experimental results of Moewus are wholly unreliable, and that the data were largely made up to fit his theories rather than based on observations. This is, in fact, now considered to be one of the very few scientific scandals in biology (see Ryan, 1955, *Science,* **122**:470).

There is no doubt about the reality of the phenomenon of relative sexuality, and it is clearly something that needs study and analysis; but the papers of Moewus have unfortunately given it unpleasant associations.

There is one special process that occurs in *Paramecium aurelia*— and rarely or not at all even in other species of the genus—that has played a large part in the genetical study of the ciliates. This process, called endomixis (or, now, autogamy) was described by Woodruff and Erdmann in 1914; their account of the cytological details was corrected and brought into better agreement with later genetic results by Diller in 1934. In this process, the macronucleus degenerates and is replaced by a division product of the micronucleus. The micronucleus itself undergoes meiosis, and one of the resulting haploid nuclei divides to give two identical nuclei—a process similar to what happens at conjugation. But at conjugation one of these two genetically identical nuclei migrates to the other individual, and there fuses with the nonmigratory product in that individual, thus producing in each ex-conjugant a diploid biparental micronucleus, from which the whole nuclear complement of its descendants arises by division. In autogamy, the two identical sister micronuclei fuse, and again the whole nuclear complement of the descendants arises by division from this uniparental diploid nucleus. It follows from this that, after autogamy has occurred, a line of *P. aurelia* individuals is genetically homozygous until conjugation (or mutation) occurs.

One result of the more recent studies is the demonstration that the phenotype of a ciliate is dependent on its macronucleus, which was derived from a micronucleus and does not survive conjugation or autogamy. The germ-plasm theory is valid here; the macronucleus is somatic, the micronucleus is germinal.

The unusual properties of the ciliates have been exploited by Sonne-born and his co-workers (Kimball, Preer, Beale, Siegel, and others—see Beale, 1954, for a summary) to produce a coherent body of knowledge about the genetics of the ciliates that is important in many current developments but is beyond the scope of this book.

CHAPTER 19

MATERNAL EFFECTS

Aristotle wrote at length on the relative effects of the male and female parents on the properties of their offspring—his discussion is now interesting as an example of the Aristotelian method rather than as a real contribution. With the increase in knowledge of development there gradually appeared the idea of preformation, according to which the fertilized egg contains the parts of the developing individual in miniature—development consisting of the unfolding of these parts in a manner similar to the development of a flower from a bud. In its extreme form, this led to the conclusion that each egg also contains miniature representatives of the eggs of all potential future descendants.

With the development of clearer ideas about fertilization, two schools emerged: the "ovists" who thought the preformed parts were contained in the unfertilized egg and were merely activated by the sperm, and the "spermists" who thought of the sperm as a complete animalcule that was merely nourished by the egg.

C. F. Wolff (1759) initiated a reaction from this view. Wolff thought of the fertilized egg as a relatively homogeneous structure, from which the parts of the embryo developed *de novo*. This view, known as epigenesis, was more in accord with the direct observations of embryologists and avoided some of the absurdities to which the preformationists had been led. It came to dominate the thinking of embryologists. It was also more easily reconciled with the cell theory and with the experimental results of the hybridizers.

The older views implied inequalities between the parents in the determination of the properties of their offspring; for example, that the form was determined by the mother, the color by the father. I have, in fact, encountered such views currently held by a few amateur plant breeders.

Kölreuter (1761–1766) seems to have been the first to carry out systematic reciprocal crosses, and to have concluded that the two parents contributed equally to the characteristics of their offspring. This conclusion was confirmed by most of the plant hybridizers who followed him (see Chapter 1), but was long resisted by zoologists.

It may be supposed that the zoologists were at first influenced by the often-discussed differences between the mule and the hinny—the results of reciprocal crosses between the horse and the ass. Hinnies are rare, and I have never been able to find a satisfactory account of them. The only supposed hinny I have ever seen impressed me as being merely a small mule—and the smaller size is the only generally recognized peculiarity. This may be due to their having a smaller mother; it is also probable that they are usually the offspring of small and inferior individuals of both parent species, since they are usually accidental in origin.

In later times zoologists were undoubtedly influenced by the study of hybrid embryos of marine animals, especially species hybrids in sea urchins. Here the effects of foreign sperm are sometimes not at once apparent, and the hybrid embryo begins its development according to the maternal plan. Often the later embryos from reciprocal crosses are not distinguishable (Boveri, 1892, 1903; Driesch, 1896, 1898). These authors concluded that up to a certain point the development is controlled entirely by the cytoplasm of the egg—though Boveri later realized that this cytoplasmic specificity might be under the control of the chromosomes of the mother. Others, however, were led to conclude that the general ground plan of development is not under chromosomal determination. As Loeb (1916, 1919) expressed it, the "embryo in the rough" is determined by the cytoplasm alone, or as Conklin (1918) stated, "we are vertebrates because our mothers were vertebrates and produced eggs of the vertebrate pattern; but the color of our skin and hair and eyes, our sex, stature and mental peculiarities were determined by the sperm as well as by the egg from which we came."

The suggestion that the maternal cytoplasm in such cases may be determined by the chromosomal genes of the mother received experimental support from the work of Toyama (1912) on the color of the embryonic serosa in the silkworm, and more clearly in the case of the snail Limnaea (Boycott and Diver, 1923; Sturtevant, 1923). In this form the shell may be coiled either dextrally or sinistrally, the two types being exact mirror images of each other. It turned out that the difference is due to a single pair of genes, with dextral dominant; but the direction of coiling is determined not by the constitution of the individual but by that of its mother. Thus, for example, a heterozygous individual, mated as a female (the animals are hermaphroditic) to another from a pure sinistral line, will produce only dextral off-spring, even though half of these individuals do not carry the gene for dextral coiling. The fate of the egg is thus determined, before polar-body formation and fertilization, by the genes of the mother. The nature of the coiling is visibly determined early in development; the two types can be distinguished by the pattern of cleavage at the second division following fertilization; but the mirror image relation persists throughout the life of the animal.

The somewhat similar case of the color of the sap in pollen grains studied by Correns has already been described in Chapter 5. Many other instances are known; for example, the gene affecting sex in *Drosophila neorepleta,* described in Chapter 13. In a few of them, the maternal nature of the hybrids is due to a failure of some or all of the paternally derived chromosomes to persist in the foreign cytoplasm (Baltzer, 1909; Godlewski, 1911).

There is another group of maternally inherited characteristics that is different in kind, namely, certain chloroplast defects in plants. In many plants there are ordinary chromosomal genes that affect the green pigment, giving regular Mendelian results, with white or pale green seedlings segregating in the usual ratios. One of these, in the snapdragon, has already been described (Chapter 8) as the first clearly demonstrated lethal gene (Baur, 1907, 1908).

In many plants there are strains in which the leaves and stems are variegated with respect to chlorophyll color; some of these behave differently. The first unambiguous case of maternal inheritance in such strains was reported by Correns (1909) in Mirabilis. In one strain of this plant the leaves are irregularly mottled dark green and yel-lowish white, the difference being in the color of the individual plas-tids. At the boundaries between the two areas there are some cells that contain both kinds of plastids. The pattern is so irregular that some branches are wholly green and others are wholly white. Correns used flowers on such uniform branches and found that seeds from those on wholly green branches gave green offspring only, regardless of the source of the pollen used; those from wholly white branches gave white seedlings only, again without regard to the source of the pollen used; flowers on variegated branches gave seeds that produced green, variegated, or white seedlings whether selfed or pollinated by wholly green plants. Evidently, then, the plastids act as though they or their precursors were self-reproducing bodies, with their properties unaf-fected by the chromosomal genes.

Similar results had been obtained by Baur (1908) with a variegated strain of Pelargonium, but this case was complicated by the transmission of some plastids through the pollen. In the earlier literature, the most instructive examples of this sort of inheritance were, perhaps, those described in Oenothera by Renner (1922, 1924). Here it was shown quite conclusively that the color of the chloroplasts is determined by the interaction of the inherent properties of the plastid precursors and of the chromosomal genes present in the individual, with each of these components maintaining and transmitting its potentialities regardless of the other, and therefore of the particular phenotype of the plant in which they occur.

Imai (1928) reviewed the results on barley that were published in Japanese by So in 1921. Here is a recessive gene for chlorophyll variegation, which evidently produces its characteristic phenotype by inducing mutations in occasional chloroplasts, causing them and the plastids descended from them to lose their green color.

A similar situation in maize was studied in detail by Rhoades (1943, 1950, and later). There is a recessive gene known as "iojap," described by Jenkins in 1924, which causes white striping in the leaves. Rhoades showed that the white plastids in the colorless areas were transmitted as such, even in the absence of the iojap gene. He had previously described a "male-sterile" line, in which the property was transmitted to all offspring of male-sterile plants when these were used as female parents; but when the small quantities of fertile pollen were used on normal plants, there were no male-sterile offspring. In 1950 he showed that this condition was regularly induced in some of the offspring of homozygous iojap plants and was then again inherited maternally even in the absence of the iojap gene.

Here then is a Mendelian recessive gene that induces permanent mutational changes in two different maternally inherited properties (the two mutations are independent, occurring in different cell lines). The male sterility is believed to depend on a mitochondrial defect, and the plastids apparently arise from mitochondria or similar bodies.

There is, then, good evidence that the plastids carry their own genes,[*] and a strong suggestion that at least some elements identified as mitochondria do so. It should be pointed out also that strictly maternal inheritance has been reported in many organisms for characters not obviously related to plastids—notably for flower size and other characters in Epilobium by Michaelis (1943 and later), for numerous characters in mosses by von Wettstein (1925 and later), and for growth rate in yeast (Ephrussi) and in Neurospora (Mitchell). Quite recently it has been found

---

[*] The germ of this idea was expressed by de Vries in 1889.

that there is DNA in plastids and in at least some mitochondria (review by Gibor and Granick, 1964, *Science* **145**: 890–897). It may therefore be supposed that these bodies carry genes of the same nature as those in the chromosomes.

A zoologist is sometimes inclined to compare the chloroplasts to the intracellular symbiotic algae found in some animals. As is well known, in *Hydra viridis* these bodies are transmitted through the eggs, and were once thought to be chloroplasts. They were supposedly separated from the host and cultured in vitro by Beijerinck (1890), who identified them as the well-known free-living green alga, *Chlorella vulgaris.* Whitney (1907) removed the algae from Hydra by treatment with glycerin and found that the alga-free individuals could be kept alive and would undergo asexual reproduction; he was, however, unable to reinfect them. Recently Siegel has reported similar results with *Paramecium bursaria* and has been able to infect alga-free lines with free-living Chlorella strains that had no known previous association with Paramecium.

Other intracellular agents have been found to be infective and also to be transmitted maternally. Apparently the first of these to be demonstrated was the organism responsible for Texas Fever in cattle. This organism is transmitted by a tick, which ingests it with the blood of an infected animal and transmits it by biting another animal. It was shown by Theobald Smith and Kilbourne (1893) that an infected female tick transmits the organism to her offspring, who can infect cattle by their first bites. Other disease-producing organisms, such as the Rickettsia of Rocky Mountain Spotted Fever in man, have since been shown to be transmitted through the eggs of ticks, rendering the offspring infective even without their having previously had any contact with an infected host.

Still other types of infective agents that are transmitted to offspring over many generations are known. Examples are: "Kappa" in Paramecium, which is responsible for the production of a substance that is toxic to uninfected animals (Sonneborn, Preer, and others); an agent responsible for CO_2 sensitivity in Drosophila (L'Heritier and others); a spirochaete in Drosophila that kills male offspring (Poulson and Malogolowkin); the "milk-substance" in mice that is transmitted from mother to offspring through the milk, and that leads to breast tumors in the adult female (Little and Bittner); and the "temperate" bacteriophages now being actively studied. This last agent forms a transition to the "infective agents" responsible for transformation and transduction in bacteria—which are too recently known for discussion here.

CHAPTER 20

THE GENETICS OF MAN

Man is, in many ways, very unsuitable as an object for the study of genetics. Families are too small for dependable determination of ratios, desired test matings cannot be made, and study of more than a very few generations for any particular purpose is not often possible. The social implications of human genetics are so great, however, that the subject *must* be investigated; and there are some real advantages in the material. For no other organism do we have such detailed and extensive information on anatomy, development, biochemistry, physiology, pathology, evolution, and population statistics. These advantages have, in fact, led to important advances in basic genetics through the study of human material, notably in connection with the blood groups (see Chapter 15) and the biochemistry of hemoglobin variants.

The systematic study of the genetics of man began before the Mendelian era with the work of Francis Galton, beginning in 1865; the two best-known accounts of his work are the books *Hereditary Genius* (1869) and *Natural Inheritance* (1889).

Following 1900 there accumulated a body of information concerning the Mendelian inheritance of a large series of aberrant conditions in man, beginning with Farabee's account of brachydactyly (short fingers) in 1905.

In 1902, Garrod and Bateson suggested that alkaptonuria is due to a single recessive gene, but the evidence did not seem conclusive until Garrod reported additional families in 1908. This case illustrates one of the difficulties in the study of the genetics of man, namely, the difficulty of finding an adequate number of critical families.

It is important that suspected cases of Mendelian inheritance in man should be recorded, so that they may be checked by other workers and, if valid, may be incorporated in studies of possible linkage and of anthropological questions. There is an unfortunate tendency, however, to accept cases as established when the evidence is so weak that it would not be considered conclusive for any organism other than man. My own experience in the field may be cited as an example. About 70 percent of people of European ancestry are able to roll up the lateral edges of the tongue, while the remaining 30 percent are unable to do so. In 1940 I suggested that this difference is due to a single pair of genes (the ability being dominant), though it was clear that a few people were able to learn to do it and that there were a few discordant pedigrees. In 1952 Matlock showed such a high frequency of discordance to exist between members of pairs of identical twins that even if an inherited component is an actuality (which is not certain), there is sufficient nongenetic influence to make the character practically useless as a genetic marker. But I am still embarrassed to see it listed in some current works as an established Mendelian case.

In spite of these difficulties, a large list of more or less clear-cut Mendelian differences in man has gradually been built up, largely concerned with relatively rare defects or with less obvious biochemical variations such as blood groups, hemoglobin types, or variations in urine composition.

These cases have been important in the understanding of the genetic components of some diseases and have also been occasionally helpful in diagnosis. Mainly for these reasons, many medical schools now have departments of medical genetics and several standard books on the subject have been published. The clear-cut cases have also been of importance in physical anthropology (the beginnings of this application were described in Chapter 15).

The more obvious and familiar human differences, such as stature, hair form and color, eye color, skin color, right– vs. left-handedness, or fingerprint patterns, although obviously inherited, are difficult to analyze. In other mammals, hair color and eye color are among the best understood of the inherited characteristics, but in man there are so many intermediates that analysis is difficult. Red hair and blue eyes are often listed as due to recessive genes, which they may be, but in both cases classification is often uncertain, and if one depends on the usual popular descriptions, there will be contradictory observations.

Even more difficult to analyze are mental properties, and obviously these are the human characteristics that are of the greatest interest and importance to society. At the sensory level, there are well-established Mendelian differences that must have indirect effects on behavior—such things as taste sensitivity, night blindness, or color blindness. Since I am partially color blind, I am acutely aware of some of the effects of my relative insensitivity to redness. Sunsets or desert colors are clearly lesser sources of esthetic satisfaction to me than they are to most people, and I am so unaware of the redness caused by inflammation that I could never have been a successful practicing physician.

At the other extreme, there are more or less clearly established Mendelian cases that involve serious mental conditions—such things as Huntington's chorea (Huntington, 1872, and many post-Mendelian references) and phenylketonuria (Fölling, 1934; Penrose, 1935; Fölling, Mohr, and Ruud, 1945).

It is the range between these extremes that is both the most interesting and the most difficult to analyze. One of the first attempts was made by Galton. He was responsible for the expression "nature vs. nurture" in the determination of human characteristics, although it is probable that he assumed his readers would recognize Shakespeare as the source of the expression (in *The Tempest*, concerning what led to Caliban's properties). Galton (1869) collected a series of pedigrees showing the concentration of particular kinds of exceptional achievements in particular families, such as musicians in the Bach family. He minimized the effect of family tradition and concluded that the results were primarily due to biological inheritance, despite one case that he pointed out but did not emphasize. In the Roman family of the Scipios there was an extraordinary concentration of generals and orators, but one of them (Scipio Aemilianus) "was not of Scipio blood" but was an adopted son, suggesting (though not to Galton) the importance of family tradition rather than genetic composition.

This same approach was later followed by Davenport *(Heredity in Relation to Eugenics,* 1911). Here there is a description of the Tuttle-Edwards family of New Haven, from whom descended two presidents and one vice-president of the United States, six college presidents, and other notables; and of the Lees of Virginia, who ran to generals and political figures. There follow accounts of the Jukes and Kallikak families, with their dreary processions of prostitutes, thieves, drunkards, and paupers. Here again was little or no recognition of the overwhelming importance of family environment and of the resulting opportunities or lack of opportunities in these examples. Surely Davenport must have understood that a potential college president, or member of the Virginia legislature, born into a Jukes family would have had no chance of realizing those potentialities—but the book does not bring out this point.

Similar views have been expressed since 1900 by other

biologists—including some who were more sophisticated than Davenport. Two examples follow:

Bateson (1912, "Biological Fact and the Structure of Society," Herbert Spencer lecture, Oxford):

> How hard it is to realize the polymorphism of man! Think of the varieties which the word denotes, merely in its application to one small society such as ours, and of the natural genetic distinctions which differentiate us into types and strains—acrobats, actors, artists, clergy, farmers, labourers, lawyers, mechanics, musicians, poets, sailors, men of science, servants, soldiers, and tradesmen. Think of the diversity of their experience of life. How few of these could have changed parts with each other. Many of these types are, even in present conditions, almost differentiated into distinct strains . . . I never cease to marvel that the more divergent castes of civilized humanity are capable of interbreeding and of producing fertile offspring from their crosses. Nothing but this paradoxical fact prevents us from regarding many classes even of Englishmen as distinct species in the full sense of the term.

Darlington (1953, *The Facts of Life*): "In England, for example, it is not lack of research which limits food production but the genetic unfitness of a large part of the tenant farmers, the legally secured occupiers who are organized to keep better men off the land."

Such extreme views have not gone unchallenged. Especially among anthropologists (largely under the influence of Boas) and among psychologists there has been a strong tendency to minimize the effects of genetic composition on human behavior. The most extreme statement of this position that I know is by Watson (1925, *Behaviorism*): "In the case of man, all healthy individuals . . . start out *equal*. Quite similar words appear in our far-famed Declaration of Independence. The signers of that document were nearer right than one might expect, considering their dense ignorance of psychology. They would have been strictly accurate had the clause 'at birth' been inserted after the word equal."

Much of the discussion of this question has been on the emotional level, because unambiguous objective evidence is so difficult to get. By and large, the extreme proponents of genetic determination have tended to be political conservatives with their views ultimately rooted in the caste system of feudalism, while the extreme advocates of environmental control have tended to represent a political philosophy derived more from the egalitarianism of the French Revolution.

As it happens, the most effective approach to this question was initiated by Galton (1883, *Inquiries into Human Faculty*). In a series of studies on pairs of twins, he recognized that they were of two kinds, "similar" and "dissimilar," and concluded that these arose, respectively, from a single fertilized egg and from two independently fertilized eggs. This conclusion has since been confirmed by embryological evidence and by extensive genetic studies; the two types are now usually referred to as monozygotic (or identical) and dizygotic (or fraternal). Galton saw that they offered an opportunity to test the relative importance of nature and nurture, since the monozygotics should be alike in genetic makeup, whereas the dizygotics should be no more alike than ordinary brothers and sisters. He carried out a few tests on mental properties and concluded that the monozygotics were in fact more alike in behavioral attributes.

The next step was taken by Muller (1925). He found a pair of monozygotic twins who had been separated in early life and brought up in different families. He gave them a series of psychological tests, and found them to be quite similar. This method was greatly extended by Newman, Freeman, and Holzinger (1937). They found a considerable series (twenty) of such separated monozygotics and, as controls, carried out the same tests on a series of monozygotics, and also of dizygotics, reared together. The book makes fascinating reading—especially the detailed case histories—but the authors admitted to disappointment at the inconclusiveness of the results. Later series of such studies have also been rather disappointing, although there can be no question of their importance. Among the difficulties encountered may be mentioned the uncertainty as to just what the psychological tests are measuring, the varying ages at which the separations took place in the different pairs, the inaccuracy of the underlying tacit assumption that twins reared together are exposed to identical environmental effects, and the circumstance that the separated twins were usually reared in rather similar families (never was one brought up as a Lee and his twin as a Jukes). Nevertheless, these studies have convinced most unbiased students that there is an appreciable inherited component in the determination of human mental differences.

The difficulties of objective study of mental differences reach their maximum in the case of racial differences. If it be admitted that there are inherited individual differences, then on general grounds one must conclude that there are statistical differences between races. If one is inclined to look upon individual mental differences as largely genetic in origin, he then is likely to consider the observed (or imagined) cultural differences between races as being genetically determined and to conclude that some races (inevitably including the one to which he belongs) are inherently superior. The extreme examples of this attitude have not usually been scientifically trained; the terrible example is Hitler, of course, but he was preceded by many pseudoscientific writers (such as Gobineau, Houston Chamberlain, and Madison Grant), most of whom would have been horrified by Hitler's methods. There have, however, been biologists with some background in genetics who have leaned in this direction. Since racism is a dirty word, it is perhaps kinder (and certainly more agreeable to the writer) not to name them.

Galton was one of the first to suggest the possibility of the genetic improvement of human populations; he introduced the word *eugenics* to designate this field of study and planning. There are two approaches here, which have been described as "negative" and "positive." The first proposes to decrease or eliminate the more extreme inherited defects—physical and mental—and the second proposes to increase the number of better individuals, and thereby to make possible the production of still better ones. Both approaches, especially the positive one, are based on the obvious success of animal and plant breeders in improving the populations with which they work.

It is estimated that something like 4 percent of human infants have tangible defects that can be detected in infancy—some of them very serious and others much less so, and some of them remediable and others not. It is also estimated that perhaps about half of these are largely genetic in origin. If it were possible to eliminate these by preventing their birth, this would obviously be a great advantage to society, in economic and, especially, in humanitarian terms.

In the early days of Mendelism, there were many people who felt that this objective could be rather simply achieved, but with increased knowledge this hope has been somewhat dimmed. The easiest class of defects to eliminate should be the dominant, but it has turned out that the more serious of these are apt not to appear until the normal reproductive age has largely passed (the typical example here is Huntington's chorea). Presumably those that appear earlier in life have, for the most part, been eliminated by natural selection. Any appreciable decrease in the incidence of recessive defects would depend on the identification of heterozygous carriers—which is not usually possible. There has also come to be a growing realization that, in some cases, heterozygosis for a particular gene may (at least under certain conditions) confer an advantage even when homozygosis is very disadvantageous. The best-known example here is sickle-cell anemia in man. Homozygosis for this gene causes the serious defect from which the name is derived; but it was shown by

Allison (1954) that heterozygosis for it confers considerable resistance to malaria and so is of selective advantage where malaria is prevalent. It remains uncertain how frequent this type of relation is, but the possibility suggests that caution be exercised in any attempt to eliminate undesirable recessives. A further point has been emphasized by Haldane, namely, that a recessive which interferes with the fertility of the individual must be retained in the population largely by recurrent mutation and therefore cannot be eliminated by artificial selection, although its frequency may be reduced.

Positive eugenics seems even more difficult, for several reasons. It is evident that animal breeders have, by selection from mixed populations, produced many reasonably uniform breeds, possessing desired characteristics and including many individuals more extreme in these respects than any found in the original population. There is no reason to doubt that similar results could be obtained with human populations. But there are a whole series of obvious difficulties—of which the greatest is: Who sets the goals? Who functions as the animal breeders have, in determining the basis of selection? Obviously no sane person would want a Hitler to have this power and responsibility, and most of us would agree with Bateson in mistrusting even a committee of Shakespeares.

CHAPTER 21

GENERAL REMARKS

There is a widespread view that scientific discoveries are more or less inevitable, and that it makes little difference whether or not a particular individual makes a discovery at a given time: if the time is not ripe for it, it will not be understood and will have little or no effect on future events; if the time is ripe, then someone else will soon make the discovery anyhow.

The history of Mendelism is one of the often-cited examples here. According to this interpretation, Mendel's paper was not understood in 1866 because the time was not ripe; in 1900, when the time was ripe, the principles were discovered independently by three different people. To me, this account seems greatly oversimplified—though it must be admitted that the development of the subject would probably have been much the same, even to the dates, if Mendel's paper had never been written.

It is true that the paper was ahead of its time, but it was not difficult to understand, and it seems unlikely that it would have remained unappreciated for so long if it had appeared in a less obscure journal, or if Mendel himself had published the further cases that he reported in his letters to Nägeli. It must be remembered that Nägeli's failure to appreciate the paper in 1866 can be matched by Pearson's failure in 1904. Both were outstanding men, and both were actively studying heredity, but to both of them Mendel's results appeared as trivial cases involving a few superficial characters, obviously neither useful nor illuminating for any general theory of inheritance. It does not follow that no biologist was likely to have appreciated the paper if he had seen it before 1900—I have suggested above that Galton, for one, would very likely have done so.

As for the simultaneous discovery in 1900, I have pointed out in Chapter 4 that it seems likely that the independent discovery was the finding of Mendel's paper, and that the actual working out of the principles without knowledge of Mendel's work was only accomplished once—by Correns—and even here it is not possible to be certain how clearly he understood the principles before he read Mendel.

In connection with the idea of the inevitability of scientific discoveries, it seems necessary to inquire into the meaning of the expression "when the time is ripe." The state of knowledge and opinion at a given time is obviously the result of individual intellectual efforts and can scarcely be thought of as inevitable. It does make a difference whether a discovery is made now, or next year, since the whole course of events in the intervening year is altered by the discovery; if it is not made now, there is a chance that the time may be overripe next year, since attention may have shifted to a quite different field.

There are other examples of a widespread failure to appreciate first-rate discoveries in genetics, and it is perhaps worthwhile to examine some of these briefly. Perhaps the most remarkable examples are the work of Cuénot on multiple alleles, of Renner on Oenothera, and of Garrod on biochemical genetics. These were all accessible and were often referred to, and all were written by men with established reputations— yet they were not fully understood nor was their importance realized until several years later. This neglect seems to have arisen in part, in all three cases, from failure to understand the terminology used. All three authors wrote in a simple, direct style, and their ideas were not inherently difficult to understand.

Cuénot used a set of symbols for the genes that was unorthodox and confusing, and he seems not to have realized that the multiple allelism that he demonstrated was unusual or unexpected. Renner was dealing with a complex situation, and he developed a useful simplifying terminology to describe it—with the result that the later beautifully clear and illuminating papers were unintelligible unless this terminology was first learned. Garrod was concerned with biochemical processes, and few geneticists were well enough grounded in biochemistry to be willing to make the moderate effort required to understand what he was talking about.

Mendel worked alone, and some of the more recent geneticists have also been rather solitary. Correns, for example, was inclined to look for new material and new problems as soon as others began to work at the problems that concerned him. Johannsen was also a rather isolated person. But it has become more and more usual for geneticists to work in closely collaborating groups, a tendency the subject shares with most scientific disciplines. The first such group was organized by Bateson, initially at Cambridge, and then at the John Innes Horticultural Institution. This group, unlike most of the later ones, used a wide variety of experimental objects—fowl, rabbits, stocks, peas, sweet peas, Primula, and many other forms. It was, however, a closely collaborating group, with much exchanging of ideas and mutual stimulation. The more recent schools have tended rather to concentrate on particular forms. A few examples are: the group organized by Castle at the Bussey Institution at Harvard, working on rodents; Emerson's maize group, at first at Nebraska but more especially later at Cornell; Morgan's Drosophila group at Columbia, and later at the California Institute of Technology with offshoots at Texas, Indiana, Columbia again, Moscow, Edinburgh, and elsewhere; Beadle's and Tatum's Neurospora group at Stanford, and many similar groups. As a rule, people working in separate laboratories on similar problems are in close contact through correspondence, temporary residence at each other's institutions, and frequent specialized symposia.

The development of genetics is one of the striking examples of the interaction between different disciplines. After 1900, the first such interaction was with cytology, which led to a very rapid development of both subjects. Later interactions were with statistics, practical breeding, evolution theory, immunology, and biochemistry. All of these have led to the utilization of new ideas and new techniques, and to rapid—sometimes spectacular—advances in genetics and in the other fields concerned.

The history of a science is primarily a history of ideas and, as such, I have treated it largely from a biographical point of view. It is also possible to treat it with emphasis on the development of new techniques—in the case of genetics, such things as the study of chromosomes, use of statistical methods, of irradiation, or of biochemical methods—or on the introduction of new kinds of organisms that are especially favorable for the study of particular problems—such as Drosophila, Neurospora, Paramecium, bacteria, or bacteriophages.

CHRONOLOGY

circa 323 B.C. ARISTOTLE: nature of reproduction and inheritance; species hybrids; recorded Drosophila.

1676 GREW: sex in plants.

1677 LEEUWENHOEK: saw animal sperm.

1716 MATHER: effects of cross-pollination in maize.

1759 WOLFF: epigenesis.

1761–1766 KÖLREUTER: began systematic study of hybrid plants.

1823–1846 AMICI: fertilization in seed plants.

1853 THURET: fertilization observed (in Fucus).

1859 DARWIN: *Origin of Species.*

1866 MENDEL: paper on peas.

1868 DARWIN: *Variation in Animals and Plants.*

1871 MIESCHER: "nuclein" (nucleoprotein).

1875 O. HERTWIG: fertilization of the sea-urchin egg.

1881 FOCKE: reference to Mendel.

1882–1885 FLEMMING, FOL, STRASBURGER, VAN BENEDEN, BOVERI, *et al.:* chromosome behavior worked out in some detail.

1883 ROUX: hypothesis on function of mitosis.

1883–1889 WEISMANN: germ-plasm theory.

1888–1889 MAUPAS: conjugation and senescence in ciliates.

1889 ALTMANN: nucleic acid.

DE VRIES: *Intracellular Pangenesis.*

1894 BATESON: *Materials for the Study of Variation.*

1900 CORRENS, DE VRIES, TSCHERMAK: rediscovery of Mendel's paper, and confirmation of his results.

LANDSTEINER: human blood groups.

1901 MCCLUNG: X chromosome as sex determinant.

DE VRIES: *Die Mutationstheorie.*

1902 BATESON, CUÉNOT: Mendelism in animals.

BOVERI: polyspermy experiments and the individuality of the chromosomes.

CORRENS: time and place of segregation.

1903 LEVENE: chemical distinction between DNA and RNA.

SUTTON: chromosomes and Mendelism.

1904 CUÉNOT: multiple alleles.

1905 BATESON AND PUNNETT: linkage.

STEVENS, WILSON: relation of sex chromosomes to sex determination.

1906 DONCASTER AND RAYNOR: sex-linkage.

LOCK: suggested the relation between linkage and exchange of parts between homologous chromosomes.

1907 E. AND E. MARCHAL, LUTZ: polyploidy.

1907–1908 BAUR: lethal gene in Antirrhinum.

1908 GARROD: alkaptonuria and genetic analysis of metabolism.

HARDY, WEINBERG: equilibrium formula for Mendelian populations.

LUTZ: trisomy.

NILSSON-EHLE: multiple gene interpretation.

1909 CORRENS: demonstration of plastid inheritance.

JANSSENS: chiasmatype hypothesis.

JOHANNSEN: *Elemente der exakte Erblichkeitslehre.*

1910 VON DUNGERN AND HIRSZFELD: heredity of human ABO blood groups.

MORGAN: sex-linkage in Drosophila; recombination between sex-linked genes.

1911 MORGAN: linkage between sex-linked genes; strength of linkage due to nearness together in a chromosome.

1912 GOLDSCHMIDT: intersexuality in Lymantria.

MORGAN: recessive lethal gene.

1913 EMERSON AND EAST: multiple genes in maize.

STURTEVANT: chromosome maps based on linkage.

1914 BRIDGES: cytology and nondisjunction.

RENNER: balanced lethals in Oenothera.

1915 MORGAN, STURTEVANT, MULLER, AND BRIDGES: *The Mechanism of Mendelian Heredity.*

1916 LITTLE AND TYZZER: genetics of susceptibility to transplanted tumors.

1917 WINGE: polyploidy.

1919 CASTLE: multiple genes and selection.

RENNER: pollen lethals in Oenothera.

1921 BRIDGES: triploidy, genic balance, and sex determination.

1922   Cleland: chromosome rings in Oenothera.

L. V. Morgan: attached-X in Drosophila.

1924   Haldane: algebraic analysis of the effects of selection.

1925   Anderson: proof of 4-strand crossing over.

Bernstein: multiple allele interpretation of human ABO blood groups.

Sturtevant: position effect.

1926   Sturtevant: genetic proof of inversion.

1927   Belling: interpretation of chromosome rings.

Landsteiner and Levine: MN blood groups in man.

Loeb and Wright: genetics of transplant specificity in mammals.

Muller: induction of mutations by X rays.

1928   Griffith: transformation in Pneumococcus.

1930   Fisher: *Genetical Theory of Natural Selection.*

Todd: blood-group specificity in fowl.

1932   Wright: genetic drift and evolution.

1933   Heitz and Bauer, Painter: nature of salivary gland chromosomes.

1935   Ephrussi and Beadle: transplantation work on Drosophila eye colors begun.

Winge: sexual reproduction in yeast.

1937   Dobzhansky: *Genetics and the Origin of Species.*

Sonneborn: mating types in Paramecium.

1940   Butenandt, Weidel, and Becker: v^+ substance is kynurenine.

Landsteiner and Wiener: Rh blood groups in man.

1941   Beadle and Tatum: biochemical mutants in Neurospora.

1944   Avery, MacLeod, and McCarty: transforming agent in Pneumococcus is DNA.

1945   Lewis: beginning of pseudoallelism study.

Owen: blood groups in cattle twins.

1946   Hershey: recombination in bacteriophage.

INTELLECTUAL
PEDIGREES

The accompanying diagrams are presented with some hesitation, since they are certainly greatly oversimplified. In general, the attempt has been to show teacher-student relationships and to neglect the interactions between contemporaries and of students on their teachers—both of which are evidently often important. They also neglect the effects of brief contacts and of influences through reading, rather than direct personal relations. Finally, they are very incomplete even for what has been attempted, and I am sure there are inaccuracies.

There nevertheless seems to be some advantage in attempting to give a picture of the various interrelations and a sense of the continuity of personal influences—even in an incomplete and imperfect form.

The information has been derived from many sources: biographies and other published accounts, my own personal knowledge, and the help and advice of numerous friends about their own backgrounds and about others with whom they were familiar.

The diagrams are self-explanatory for the most part. It will be seen that the same name often occurs in several; this is because of an effort to keep them in a relatively simple form. In one case (p. 141, top) there is a series of names enclosed in parentheses; these represent men who came as postdoctoral students, already trained elsewhere, to the California Institute of Technology where they were influenced by the group shown.

Many names that might well have been included are omitted in order to simplify the diagrams; I have had to be quite arbitrary about this.

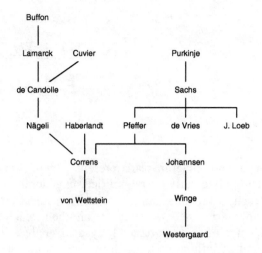

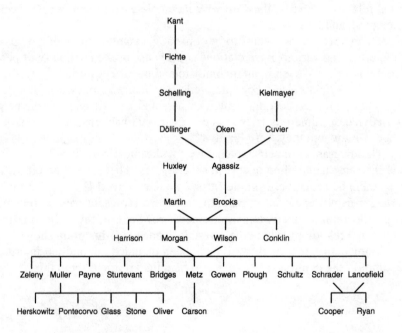

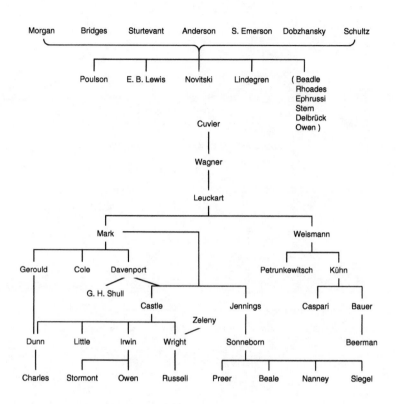

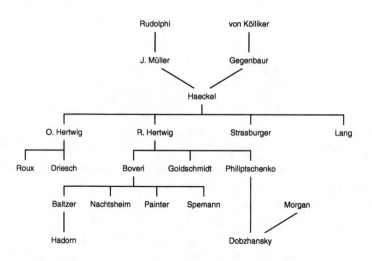

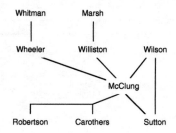

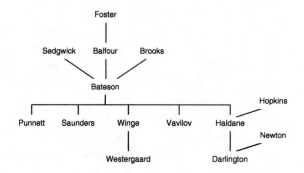

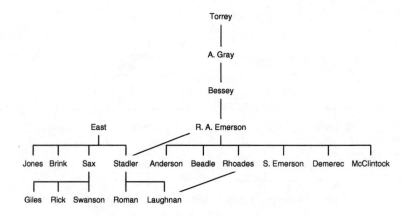

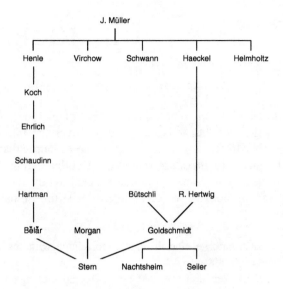

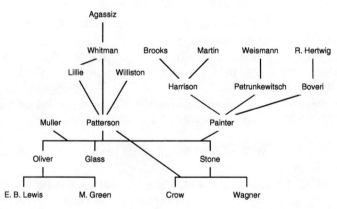

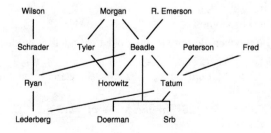

BIBLIOGRAPHY

The list of publications that follows does not include all the references given in the text, but it is designed to cover the most important papers in the mainstream of genetics, plus some accounts that are themselves useful sources of literature citations. A fuller listing might have been useful to the serious student, but that seemed to be out of proportion with the rest of this book. The selection made is, unfortunately, somewhat arbitrary.

Allen, C. E. 1917. A chromosome difference correlated with sex in Sphaerocarpos. *Science,* 46: 466–467.

Allison, A. C. 1954. Protection afforded by sickle-cell trait against subtertian malarial infection. *Brit. Med. J.* (4857): 290–294.

Anderson, E. G. 1925. Crossing over in a case of attached X chromosomes in *Drosophila melanogaster. Genetics,* 10: 403–417.

Aristotle. *Historia Animalium.* Translated by D'Arcy W. Thompson. 1910. Clarendon Press, Oxford (vol. 4, *The Works of Aristotle*). Original work, about 323 B.C.

Aristotle. *De Generatione Animalium.* Translated by A. Platt. 1912. Clarendon Press, Oxford (vol. 5, *The Works of Aristotle*). Original work, about 323 B.C.

Auerbach, C., and J. M. Robson. 1946. Chemical production of mutations. *Nature,* 157: 302.

Avery, O. T., C. M. MacLeod, and M. McCarty. 1944. Studies on the chemical nature of the substance inducing transformation of pneumococcal types. *J. Exper. Med.* 79: 137–158.

Balbiani, E. G. 1881. Sur la structure du noyau des cellules salivaires chez les larves de Chironomus. *Zool. Anz.,* 4: 637–641, 662–666.

Baltzer, F. 1909. Über die Entwicklung der Echiniden-bastarde. *Zeits. ind. Abst. Vererb.,* 5.

Bateson, B. 1928. *William Bateson, Naturalist.* Cambridge Press, Cambridge. 473 pp.

Bateson, W. 1894. *Materials for the Study of Variation.* The Macmillan Co., New York. 598 pp.

Bateson, W. 1902. *Mendel's Principles of Heredity: A Defence.* University Press, Cambridge. 212 pp.

Bateson, W. 1909. *Mendel's Principles of Heredity.* University Press, Cambridge. 396 pp.

Bateson, W. 1916. The mechanism of Mendelian heredity (a review). *Science,* 44: 536–543.

Bateson, W., and E. R. Saunders. 1902. Experimental studies in the physiology of heredity. *Reports to Evol. Comm. Royal Soc.,* 1. 160 pp.

Bateson, W., E. R. Saunders, and R. C. Punnett. 1905. Experimental studies in the physiology of heredity. *Reports to Evol. Comm. Royal Soc.,* 2: 1–131.

Bateson, W., E. R. Saunders, and R. C. Punnett. 1906. Experimental studies in the physiology of heredity. *Reports to Evol. Comm. Royal Soc.,* 3: 1–53.

Baur, E. 1908. Untersuchungen über die Erblichkeitsverhältnisse einer nur in Bastardform lebensfähigen Sippe von *Antirrhinum majus. Zeits. ind. Abst. Vererb.,* 1: 124.

Beadle, G. W., and S. Emerson. 1935. Further studies of crossing over in attached-X chromosomes of *Drosophila melanogaster. Genetics,* 20: 192–206.

Beadle, G. W., and E. L. Tatum. 1941. Genetic control of biochemical reactions in Neurospora. *Proc. Nat. Acad. Sci.,* 27: 499–506.

Beale, G. H. 1954. *The Genetics of Paramecium aurelia.* Cambridge Press, Cambridge. 179 pp.

Belling, J. 1927. The attachment of chromosomes at the reduction division in flowering plants. *J. Genet.,* 18: 177–205.

Belling, J. 1931. Chromomeres of lilaceous plants. *Univ. Calif. Publ. Botany,* 16: 153–170.

Belling, J., and A. F. Blakeslee. 1924. The configurations and sizes of the chromosomes in the trivalents of 25-chromosome Daturas. *Proc. Nat. Acad. Sci.,* 10: 116–120.

Benzer, S. 1961. On the topography of the genetic fine structure. *Proc. Nat. Acad. Sci.,* 47: 403–415.

Bernstein, F. 1925. Zusammenfassende Betrachtungen über die erblichen Blutstrukturen des Menschen. *Zts. Abst. Vererb.,* 37: 237–270.

Blakeslee, A. F., and A. G. Avery. 1937. Methods of inducing doubling of chromosomes in plants. *J. Hered.,* 28: 392–411.

Bostian, C. H. 1935. Multiple alleles and sex determination in Habrobracon. *Genetics,* 24: 770–776.

Boveri, T. 1902. Über mehrpolige Mitosen als Mittel zur Analyse des Zellkerns. *Verh. phys.-med. Gesellsch. Würzburg,* 35: 67–90.

Boveri, T. 1904. *Ergebnisse über die Konstitution der chromatischen Substanz des Zellkerns.* Fischer, Jena. 130 pp.

Boycott, A. E., and C. Diver. 1923. On the inheritance of sinistrality in *Limnaea peregra. Proc. Roy. Soc.,* 95B: 207–213.

Bridges, C. B. 1913. Nondisjunction of the sex chromosomes of Drosophila. *J. Exper. Zool.,* 15: 587–606.

Bridges, C. B. 1916. Nondisjunction as proof of the chromosome theory of heredity. *Genetics,* 1: 1–52, 107–163.

Bridges, C. B. 1921. Genetical and cytological proof of nondisjunction of the fourth chromosome of *Drosophila melanogaster. Proc. Nat. Acad. Sci.,* 7: 186–192.

Bridges, C. B. 1921. Triploid intersexes in *Drosophila melanogaster. Science,* 54: 252–254.

Bridges, C. B. 1925. Sex in relation to chromosomes and genes. *Amer. Nat.,* 59: 127–137.

Bridges, C. B. 1935. Salivary chromosome maps. *J. Hered.,* 26: 60–64.

Bridges, C. B. 1938. A revised map of the salivary gland X-chromosome. *J. Hered.,* 29: 11–13.

Bridges, C. B. and P. N. Bridges. 1939. A new map of the second chromosome. *J. Hered.,* 30: 475–477.

Bridges, C. B., and T. H. Morgan. 1919. The second-chromosome group of mutant characters. *Carn. Inst. Wash.,* publ. 278: 123–304.

Bridges, C. B., and T. H. Morgan. 1923. The third-chromosome group of mutant characters of *Drosophila melanogaster. Carn. Inst. Wash.,* publ. 327, 251 pp.

Butenandt, A., W. Weidel, and E. Becker. 1940. Kynurenin als Augenpigment-bildung auslösendes Agens bei Insekten. *NaturWiss.,* 28: 63–64.

Carothers, E. E. 1913. The Mendelian ratio in relation to certain Orthopteran chromosomes. *J. Morph.,* 24: 487–511.

Caspari, E. 1933. Über die Wirkung eines pleiotropen Gens bei der Mehlmotte Ephestia Kühniella. *Arch. Entw.-mech.,* 130: 352–381.

Castle, W. E. 1903. The laws of heredity of Galton and Mendel and some laws governing race improvement by selection. *Proc. Amer. Acad. Arts Sci.,* 39: 223–242.

Catcheside, D. G. 1947. The *P*-locus position effect in Oenothera. *J. Genet.,* 48: 31–42.

Chetverikov, S. S. 1926. (Russian, text in *Zhurn. Exp. Bio., A2: 3–54.)* Translated, 1961. On certain aspects of the evolutionary process from the standpoint of modern genetics. *Proc. Amer. Philosoph. Soc.,* 105: 167–195.

Clausen, R. E., and T. H. Goodspeed. 1925. Interspecific hybridization in Nicotiana. II. *Genetics,* 10: 278–284.

Cleland, R. E. 1923. Chromosome arrangements during meiosis in certain Oenotheras. *Amer. Nat.,* 57: 562–566.

Cleland, R. E., and A. F. Blakeslee. 1931. Segmental interchange, the basis of chromosomal attachments in Oenothera. *Cytologia,* 2: 175–233.

Correns, C. 1900. Untersuchungen über die Xenien bei Zea Mays. *Ber. deutsch. botan. Gesellsch.,* 17: 410–418.

Correns, C. 1900. G. Mendels Regel über das Verhalten der Nachkommenschaft der Rassenbastarde. *Ber. deutsch. botan. Gesellsch.,* 18: 158–168. Translated, 1950, in The Birth of Genetics. Suppl. *Genetics,* 33–41.

Correns, C. 1902. Über den Modus unde den Zeitpunkt der Spaltung der Anlagen bei den Bastarden vom Erbsen-Typus. *Botan. Zeitg.,* 60, II, 5/6: 65–82.

Correns, C. 1905. Gregor Mendels Briefe an Carl Nägeli, 1866–1873. *Abh. math-phys. Kl. Kön. Sächs. Gesellsch. Wiss.,* 29: 189–265. Reprinted in *Carl Correns, Gesammelte Abhandlungen.* F. von Wettstein, ed., Springer, Berlin. Translated, 1950, in The Birth of Genetics. Suppl. *Genetics,* 1–29.

Correns, C. 1909. Vererbungsversuche mit blass (gelb) grünen und buntblättrigen Sippen bei Mirabilis, Urtica, und Lunaria. *Zeits. ind. Abst. Vererb.,* 1: 291–329.

Correns, C. 1910. Der Übergang aus dem homozygotischen in einen heterozygotischen Zustand im selben Individuum bei buntblättrigen und gestreiftblühenden Mirabilis-Sippen. *Ber. deutsch. botan. Gesellsch.,* 28: 418–434.

Correns, C. 1922. Etwas über Gregor Mendels Leben und Wirken. *Die Naturwiss,* 10: 623–631.

Creighton, H. B., and B. McClintock. 1931. A correlation of cytological and genetical crossing-over in Zea Mays. *Proc. Nat. Acad. Sci.,* 17: 485–497.

Crew, F. A. E., and R. Lamy. 1935. Linkage groups in Drosophila pseudoobscura. *J. Genet.,* 30: 15–29.

Cuénot, L. 1902. La loi de Mendel et l'hérédité de la pigmentation chez les souris. *Arch. zool. expér. gén.,* 3d series, 10, notes et revue, pp. xxvii–xxx.

Cuénot, L. 1904. L'hérédité de la pigmentation chez les souris. *Arch. zool. expér. gén.,* 4th series, 2, notes et revue, pp. xlv. lvi.

Cuénot, L. 1905. Les races pures et leurs combinaisons chez les souris. *Arch. zool. Expér. gén.,* 4th series, 3, notes et revue, pp. cxxiii–cxxxii.

Cuénot, L. 1907. L'hérédité de la pigmentation chez les souris. *Arch. zool. expér. gén.,* 4th series, 6, notes et revue, pp. i–xiii.

Darlington, C. D. 1932. *Recent Advances in Cytology.* London. 559 pp.

Darwin, C. 1868. *The Variation of Animals and Plants under Domestication.* (I have used the second edition, of 1876. 2 vols., 473 pp. and 495 pp. Appleton, New York.)

Darwin, C. 1876. *The Effects of Cross and Self Fertilization in the Vegetable Kingdom.* (1892 edition, 482 pp. Appleton, New York.)

Demerec, M. 1934. Biological action of small deficiencies of X-chromosome of *Drosophila melanogaster*. *Proc. Nat. Acad. Sci.,* 20: 354–359.

Demerec, M. 1940. Genetic behavior of euchromatic segments inserted into heterochromatin. *Genetics,* 25: 618–627.

Diller, W. F. 1934. Autogamy in a *Paramecium aurelia*. *Science,* 79: 57.

Dobzhansky, T. 1929. Genetical and cytological proof of translocations involving the third and the fourth chromosomes of *Drosophila melanogaster*. *Biol. Zentralbl.,* 49: 408–419.

Dobzhansky, T. 1930. Cytological map of the second chromosome of *Drosophila melanogaster*. *Biol. Zentralbl.,* 50: 671–685.

Dobzhansky, T. 1937. *Genetics and the Origin of Species.* Second edition (1941), Columbia University Press, New York. 446 pp.

Dobzhansky, T., and J. Schultz. 1934. The distribution of sex-factors in the X-chromosome of *Drosophila melanogaster*. *J. Genet.,* 28: 349–386.

Doncaster, L., and G. H. Raynor. 1906. Breeding experiments with Lepidoptera. *Proc. Zool. Soc. London,* 1: 125–133.

Dubinin, N. P., and B. N. Sidoroff. 1934. Relation between the effect of a gene and its position in the system. *Amer. Nat.,* 68: 377–381.

Dubinin, N. T., *et al.* 1936. Genetic constitution and gene-dynamics of wild populations of *Drosophila melanogaster*. *Biol. Zhurn.,* 5: 939.

Dungern, E. von, and L. Hirszfeld. 1910. Ueber Vererbung gruppenspezifischer Strukturen des Blutes. *Zts. Immunforsch.,* 6: 284–292.

Dunn, L. C., editor. 1951. *Genetics in the Twentieth Century.* 634 pp. The Macmillan Co., New York.

Emerson, R. A. 1914. The inheritance of a recurring somatic variation in variegated ears of maize. *Res. Bull. Agric. Exper. Sta. Nebr.,* 4: 1–35.

Emerson, R. A., and E. M. East. 1913. The inheritance of quantitative characters in maize. *Bull. Agric. Exper. Sta. Nebr.* 120 pp.

Emerson, S. 1938. The genetics of self-incompatibility in Oenothema organensis. *Genetics,* 23: 190–202.

Emerson, S., and G. W. Beadle. 1933. Crossing over near the spindle fiber in attached X chromosomes of *Drosophila melanogaster*. *Zeits. ind. Abst. Vererb.,* 65: 129–140.

Ephrussi, B. 1942. Chemistry of "eye-color hormones" of Drosophila. *Quart. Rev. Biol.,* 17: 327–338.

Ephrussi, B., and G. W. Beadle. 1935. La transplantation des disques imaginaux chez le Drosophile. *C. R. Acad. Sci.* (Paris) 201: 98.

Federley, H. 1913. Das Verhalten der Chromosomen bei der Spermatogenese der Schmetterlinge *Pygaera anachoreta*, curtula und pigra sowie einiger ihrer Bastarde. *Zts. ind. Abst. Vererb.,* 9: 1–110.

Fisher, R. A. 1928. The possible modification of the response of the wild type to recurrent mutations. *Amer. Nat.,* 62: 115–126.

Fisher, R. A. 1930. *The Genetical Theory of Natural Selection.* Clarendon Press, Oxford. 272 pp.

Fisher, R. A. 1936. Has Mendel's work been rediscovered? *Ann. Sci.* 1: 115–137.

Focke, W. O. 1881. *Die Pflanzenmischlinge.* 570 pp. Borntraeger, Berlin.

Gabriel, M. L., and S. Fogel. 1955. *Great Experiments in Biology.* Prentice-Hall, Englewood Cliffs, N.J. 317 pp.

Galton, F. 1869. *Hereditary Genius.* Macmillan & Co., London. (Second edition 1892.)

Galton, F. 1883. *Inquiries into Human Faculty.* Macmillan & Co., London.

Galton, F. 1889. *Natural Inheritance.* Macmillan & Co., London. 259 pp.

Galton, F. 1908. *Memories of My Life.* Methuen & Co., London. 339 pp.

Garrod, A. E. 1908. Inborn errors of metabolism. *Lancet,* 2: 1–7, 73–79, 142–148, 214–220. Also separately published as a book, 1909. Oxford University Press, London. 168 pp.

Glass, B. 1947. Maupertuis and the beginnings of genetics. *Quart. Rev. Biol.,* 22: 196–210.

Goldschmidt, R. 1912. Erblichkeitsstudien an Schmetterlingen, I. *Zeits. ind. Abst. Vererb.,* 7: 1–62.

Goldschmidt, R. 1933. Lymantria. *Bibliogr. Genet.,* 11: 1–186.

Gowen, J. W., and E. H. Gay. 1933. Eversporting as a function of the Y-chromosome in *Drosophila Melanogaster. Proc. Nat. Acad. Sci.,* 19: 122–126.

Gowen, J. W., and E. H. Gay. 1933. Effect of temperature on eversporting eye color in *Drosophila melanogaster. Science,* 77: 312.

Green, M. M., and K. C. Green. 1949. Crossing-over between alleles at the lozenge locus in *Drosophila melanogaster. Proc. Nat. Acad. Sci.* 35: 586–591.

Griffith, F. 1928. The significance of Pneumococcal types. *J. Hyg.,* 27: 113–159.

Guyer, M. F. 1902. Some notes on hybridism, variation, and irregularities in the division of the germ-cell. *Science,* 15: 530–531.

Guyer, M. F. 1903. The germ-cell and the results of Mendel. *Cincinnati Lancet-Clinic.* 2 pp.

Haacke, W. 1893. Die Träger der Vererbung. *Biol. Centralbl.,* 13: 525–542.

Haldane, J. B. S. 1924. A mathematical theory of natural and artificial selection. *Proc. Cambridge Phil. Soc.,* 23: 19–41, 158–163, 363–372, 607–615, 838–844.

Hanson, F. B., and F. Heys. 1929. An analysis of the effects of the different rays of radium in producing lethal mutations in Drosophila. *Amer. Nat.* 63: 201–213.

Harland, S. C. 1936. The genetical conception of the species. *Biol. Rev.*, 11: 83–112.

Heitz, E. 1933. Über totale und partielle somatische Heteropycnose, sowie strukturelle Geschlechtschromosomen bei *Drosophila funebris*. *Zeits. Zellf. mikr. Anat.* 19: 720–742.

Heitz, E., and H. Bauer. 1933. Beweise für die Chromosomennatur der Kernschleifen in den Knäuelkernen von *Bibio hortulanus*. *Zeits. Zellf. mikr. Anat.* 17: 67–82.

Henking, H. 1891. Über Spermatogenese bei *Pyrrhocoris apterus*. *Zeits. wissensch. Zool.*, 51: 685–736.

Hershey, A. D. 1946. Spontaneous mutations in bacterial viruses. *C. Spr. Harb. Sympos. Quant. Biol.*, 11: 67–77.

Hirszfeld, L., and H. 1919. Serological differences between the blood of different races. *Lancet* 1919 (2): 675–679.

Iltis, H. 1924. *Gregor Johann Mendel, Leben, Werk und Wirkung*. Springer, Berlin. Translated, E. and C. Paul, 1932 (title of translation, *Life of Mendel*). W. W. Norton & Company, New York. 336 pp.

Imai, Y. 1928. A consideration of variegation. *Genetics,* 13: 544–562.

Irwin, M. R., and L. J. Cole. 1936. Immunogenetic studies of species and of species hybrids in doves. *J. Exper. Zool.,* 73: 85–108.

Janssens, F. A. 1909. La théorie de la chiasmatypie. *La Cellule,* 25: 389–411.

Jennings, H. S. 1929. Genetics of the Protozoa. *Bibliogr. Genet.,* 5: 105–330.

Johannsen, W. 1909. *Elemente der exakten Erblichkeitslehre*. Fischer, Jena. 516 pp.

Kihara, H., and T. Ono. 1926. Chromomenzahlen und systematische Gruppierung der Rumex-Arten. *Zts. Zellf. mikr. Anat.,* 4: 475–481.

Lancefield, D. E. 1922. Linkage relations of the sex-linked characters in *Drosophila obscura*. *Genetics,* 7: 335–384.

Landsteiner, K. 1900. Zur Kenntnis der antifermentativen, lytischen und agglutinativen Wirkungen des Blutserums und der Lymphe. *Zbl. Bakt.,* 27: 357–362.

Landsteiner, K., and P. Levine. 1927. Further observation on individual differences of human blood. *Proc. Soc. Exper. Biol. Med.,* 24: 941–942.

Landsteiner, K., and A. S. Wiener. 1940. An agglutinable factor in human blood recognized by immune sera for rhesus blood. *Proc. Soc. Exper. Biol. Med.,* 43: 223–224.

Lewis, E. B. 1945. The relation of repeats to position effects in *Drosophila melanogaster*. *Genetics,* 30: 137–166.

Lewis, E. B. 1950. The phenomenon of position effect. *Advances in Genetics,* 3: 73–115.

Li, C. C. 1955. *Population Genetics.* University of Chicago Press, Chicago. 366 pp.

Little, C. C. 1915. A note on multiple allelomorphs in mice. *Amer. Nat.,* 49: 122–125.

Little, C. C., and E. E. Tyzzer. 1916. Further experimental studies on the inheritance of susceptibility of a transplantable tumor. *J. Med. Res.,* 33: 393–453.

Lock, R. H. 1906. *Recent Progress in the Study of Variation, Heredity, and Evolution.* E. P. Dutton & Co., New York. 299 pp.

Loeb, J., and F. W. Bancroft. 1911. Some experiments on the production of mutants in Drosophila. *Science,* 33: 781–783.

Loeb, L., and S. Wright. 1927. Transplantation and individuality differentials in inbred families of guinea pigs. *Amer. J. Pathol.,* 3: 251–283.

Lutz, A. M. 1907. A preliminary note on the chromosomes of *Oenothera Lamarckiana* and one of its mutants, *O. gigas. Science,* 26: 151–152.

Lutz, A. M. 1909. Notes on the first generation hybrid of *Oenothera lata* and *O. gigas. Science,* 29: 263–267.

McClung, C. E. 1901. Notes on the accessory chromosome. *Anat. Anz.,* 20: 220–226.

McKusick, V. A. 1960. Walter S. Sutton and the physical basis of Mendelism. *Bull. Hist. Med.,* 34: 487–497.

Marchal, E., and E. 1906. Recherches experimentales sur la sexualité des spores chez les mousses dioiques. *Mem. Couronnes, Cl. Sci.,* Dec. 15. 50 pp.

Matlock, P. 1952. Identical twins discordant in tongue-rolling. *J. Hered.,* 43: 24.

Mavor, J. W. 1921. On the elimination of the X-chromosome from the egg of *Drosophila melanogaster* by X-rays. *Science,* 54: 277–279.

Mavor, J. W. 1923. An effect of X-rays on crossing over in Drosophila. *Proc. Soc. Exper. Biol. Med.,* 20: 335–338.

May, H. G. 1917. Selection for facet numbers in the bar-eyed race of Drosophila and the appearance of reverse mutations. *Biol. Bull.,* 33: 361–395.

Mendel, G. 1866. Versuche über Pflanzenhybriden. *Verh. naturforsch. Ver. Brünn,* 4: 3–47. Reprinted in 1901. *Flora,* 89; and 1901. *Ostwald's Klassik. exakt. Wissensch.,* no. 121. Translations: 1901. *J. Roy. Hort. Soc.,* 26; and in Bateson, W. 1909, 1913. *Mendel's Principles of Heredity;* Gabriel & Fogel 1955; Peters, J. A. 1959; and separately 1938, Harvard University Press, Cambridge, Mass. 41 pp.

Metz, C. W. 1914. Chromosome studies in the Diptera. I. *J. Exper. Zool.,* 17: 45–58.

Metz, C. W. 1938. Chromosome behavior, inheritance and sex determination in Sciara. *Amer. Nat.,* 72: 485–520.

Miescher, F. 1871. Ueber die chemische Zusammensetzung der Eiterzellen. *Hoppe-Seylers med.-chem. Untersuch.,* 441.

Montgomery, T. H. 1901. A study of the chromosomes of the germ cells of the Metazoa. *Trans. Amer. Phil. Soc.*, 20: 154–236.

Morgan, L. V. 1922. Non-criss-cross inheritance in *Drosophila melanogaster*. *Biol. Bull.*, 42: 267–274.

Morgan, T. H. 1903. *Evolution and Adaptation*. The Macmillan Co., New York. 470 pp.

Morgan, T. H. 1910. Sex limited inheritance in Drosophila. *Science*, 32: 120–122.

Morgan, T. H. 1910. The method of inheritance of two sex-limited characters in the same animal. *Proc. Soc. Exper. Biol. Med.*, 8: 17–19.

Morgan, T. H. 1911. The influence of heredity and of environment in determining the coat colors of mice. *Ann. N.Y. Acad. Sci.*, 21: 87–117.

Morgan, T. H. 1911. An attempt to analyze the constitution of the chromosomes on the basis of sex-limited inheritance in Drosophila. *J. Exper. Zool.*, 11: 365–412.

Morgan, T. H. 1912. The explanation of a new sex ratio in Drosophila. *Science*, 36: 718–719.

Morgan, T. H., and C. B. Bridges. 1919. The origin of gynandromorphs. *Carn. Inst. Wash.*, publ. 278: 1–22.

Morgan, T. H., and C. J. Lynch. 1912. The linkage of two factors in Drosophila that are not sex-linked. *Biol. Bull.*, 23: 174–182.

Morgan, T. H., A. H. Sturtevant, H. J. Muller, and C. B. Bridges. 1915. *The Mechanism of Mendelian heredity*. Henry Holt and Co., New York. 258 pp.

Muller, H. J. 1914. A factor for the fourth chromosome of Drosophila. *Science*, 39: 906.

Muller, H. J. 1918. Genetic variability, twin hybrids and constant hybrids, in a case of balanced lethal factors. *Genetics*, 3: 422–499.

Muller, H. J. 1927. Artificial transmutation of the gene. *Science*, 46: 84–87.

Muller, H. J., and E. Altenburg. 1919. The rate of change of hereditary factors in Drosophila. *Proc. Soc. Exper. Biol. Med.*, 17: 10–14.

Muller, H. J. 1925. Mental traits and heredity as studied in a case of identical twins reared apart. *J. Hered.*, 16: 433–448.

Muller, H. J., A. A. Prokofyeva-Belgovskaya, and K. V. Kossikov. 1936. Unequal crossing over in the bar mutant as a result of duplication of a minute chromosome section. *C. R. (Dokl.) Acad. Sci. U.R.S.S.*, 1(10): 83–88.

Newman, H. H., F. N. Freeman, and K. J. Holzinger. 1937. *Twins: A Study of Heredity and Environment*. The University of Chicago Press, Chicago. 369 pp.

Nilsson-Ehle, H. 1909. Kreuzungsuntersuchungen an Hafer und Weizen. *Lunds Universit. Arsskr. N.F. 5*, 2: 1–122.

Oliver, C. P. 1930. The effect of varying the duration of X-ray treatment upon the frequency of mutation. *Science,* 71: 44–46.

Owen, R. D. 1945. Immunogenetic consequence of vascular anastomosis between bovine twins. *Science,* 102: 400–401.

Painter, T. S. 1933. A new method for the study of chromosome rearrangements and the plotting of chromosome maps. *Science,* 78: 585–586.

Painter, T. S., and H. J. Muller. 1929. Parallel cytology and genetics of induced translocations and deletions in Drosophila. *J. Hered.,* 20: 287–298.

Pavan, C., and M. E. Breuer. 1952. Polytene chromosomes in different tissues of Rhynchosciara. *J. Hered.,* 43: 151–157.

Payne, F. 1918. An experiment to test the nature of the variations on which selection acts. *Indiana Univ. Studies,* 5: 1–45.

Peters, J. A. 1959. *Classic Papers in Genetics.* Prentice-Hall, Englewood Cliffs, N.J. 282 pp.

Renner, O. 1917. Die tauben Samen der Oenotheren. *Ber. deutsch. botan. Gesellsch.,* 34: 858–869.

Renner, O. 1919. Zur Biologie und Morphologie der männlichen Haplonten einiger Oenotheren. *Zeits. Bot.,* 11: 305–380.

Renner, O. 1921. Heterogametie im weiblichen Geschlecht und Embryosackentwicklung bei den Oenotheren. *Zeits. Bot.,* 13: 609–621.

Renner, O. 1924. Die Scheckung der Oenotherenbastarde. *Biol. Zentralbl.,* 44: 309–336.

Renner, O. 1925. Untersuchungen über die faktorielle Konstitution einiger komplexheterozygotischen Oenotheren. *Biblioth. Genet.,* 9, 168 pp.

Rhoades, M. M. 1943. Genic induction of an inherited cytoplasmic difference. *Proc. Nat. Acad. Sci.,* 29: 327–329.

Roberts, H. F. 1929. *Plant Hybridization before Mendel.* Princeton University Press, Princeton, N.J. 374 pp.

Robertson, W. R. B. 1916. Chromosome studies. I. *J. Morph.,* 27: 179–331.

Roux, W. 1883. *Über die Bedeutung der Kerntheilungsfiguren.* Engelmann, Leipzig, 19 pp.

Sachs, J. 1875. *History of Botany.* Translated by H. E. F. Garnsey 1890. Oxford University Press, Oxford, 568 pp.

Schultz, J. 1936. Variegation in Drosophila and the inert chromosome regions. *Proc. Nat. Acad. Sci.,* 22: 27–33.

Seiler, J. 1913. Das Verhaltung der Geschlechtschromosomen bei Lepidopteren. *Zool. Anz.,* 41: 246–251.

Smith, T., and F. L. Kilborne. 1893. Investigations into the nature, causation and prevention of southern cattle fever. *U. S. Bur. Anim. Indust. Bull. 1.* 301 pp.

Snyder, L. H. 1929. *Blood Grouping in Relation to Clinical and Legal Medicine.* The Williams & Wilkins Co., Baltimore. 153 pp.

Sonneborn, T. M. 1938. Mating types in *Paramecium aurelia. Proc. Amer. Philosoph. Soc.,* 79: 411–434.

Stadler, L. J. 1928. Mutations in barley induced by X-rays and radium. *Science,* 68: 186–187.

Stadler, L. J. 1928. The rate of induced mutation in relation to dormancy, temperature and dose. *Anat. Record,* 41: 97.

Steiner, E. 1956. New aspects of the balanced lethal mechanism in Oenothera. *Genetics,* 41: 486–500.

Stern, C. 1926. Vererbung im Y-Chromosom von *Drosophila melanogaster. Biol. Zentralbl.,* 46: 344–348.

Stern, C. 1929. Untersuchungen über Aberrationen des Y-Chromosoms von *Drosophila melanogaster. Zeits. ind. Abst. Vererb.,* 51: 253–353.

Stern, C. 1931. Zytologisch-genetische Untersuchungen als Beweise für die Morgansche Theorie des Faktorenaustauschs. *Biol. Zentralbl.,* 51: 547–587.

Stern, C. 1934. Crossing-over and segregation in somatic cells of *Drosophila melanogaster. Amer. Nat.,* 68: 164–165.

Stevens, N. M. 1905. Studies in spermatogenesis with especial reference to the "accessory chromosome." *Carn. Inst. Wash.,* publ. 36, 33 pp.

Stevens, N. M. 1908. A Study of the germ cells of certain Diptera. *J. Exper. Zool.,* 5: 359–379.

Stomps, T. J. 1954. On the rediscovery of Mendel's work by Hugo de Vries. *J. Hered.,* 45: 293–294.

Sturtevant, A. H. 1913. The linear arrangement of six sex-linked factors in Drosophila, as shown by their mode of association. *J. Exper. Zool.,* 14: 43–59.

Sturtevant, A. H. 1913. The Himalayan rabbit case, with some considerations on multiple allelomorphs. *Amer. Nat.,* 47: 234–238.

Sturtevant, A. H. 1913. A third group of linked genes in *Drosophila ampelophila. Science,* 37: 990–992.

Sturtevant, A. H. 1917. Genetic factors affecting the strength of linkage in Drosophila. *Proc. Nat. Acad. Sci.,* 3: 555–558.

Sturtevant, A. H. 1920. The vermilion gene and gynandromorphism. *Proc. Soc. Exper. Biol. Med.,* 17: 70–71.

Sturtevant, A. H. 1925. The effects of unequal crossing over at the bar locus in Drosophila. *Genetics,* 10: 117–147.

Sturtevant, A. H. 1926. A crossover reducer in *Drosophila melanogaster* due to inversion of a section of the third chromosome. *Biol. Zentralbl.,* 46: 697–702.

Sturtevant, A. H. 1948. The evolution and function of genes. *Amer. Sci.,* 36: 225–236. Reprinted in *Science in Progress* 6: 250–256 (1949) and in *Smithsonian Report for 1948,* pp. 293–304.

Sturtevant, A. H. 1959. Thomas Hunt Morgan, 1866–1945. *Biograph. memoirs Nat. Acad. Sci.,* 33: 283–325.

Sturtevant, A. H., and G. W. Beadle. 1939. *An Introduction to Genetics.* Saunders, Philadelphia. 391 pp.

Sturtevant, A. H., and E. Novitski. 1941. The homologies of the chromosome elements in Drosophila. *Genetics,* 26: 517–541.

Sturtevant, A. H., and C. C. Tan. 1937. The comparative genetics of *Drosophila pseudoobscura* and *D. melanogaster.* J. *Genet.,* 34: 415–432.

Sutton, W. S. 1902. On the morphology of the chromosome group in *Brachystola magna. Biol. Bull.,* 4: 24–39.

Sutton, W. S. 1903. The chromosomes in heredity. *Biol. Bull.,* 4: 231–251.

Todd, C. 1930. Cellular individuality in the higher animals, with special reference to the individuality of the red blood corpuscle. *Proc. Roy. Soc.,* 106 B, 20–44.

Tschermak, E. 1900. Ueber künstliche Kreuzung bei *Pisum sativum. Zeits. landw. Versuchsw. Oesterr.,* 3, Heft 5.

Tschermak, E. 1900. Über künstliche Kreuzung bei *Pisum sativum. Ber. deutsch. botan. Gesellsch.,* 18: 232–239. Translation. 1950. The Birth of Genetics. Suppl. *Genetics.* 42–47.

Ullerich, F.-H. 1963. Geschlechtschromosomen und Geschlechtsbestimmung bei einigen Calliphorinen. *Chromosoma,* 14: 45–110.

Vries, H. de. 1889. *Intracellulare Pangenesis.* 149 pp. Fischer, Jena. (Translation by C. S. Gager, 1910. Open Court, Chicago.)

Vries, H. de. 1900. Sur la loi de disjonction des hybrides. *C. R. Acad. Sci.* (Paris), 130: 845–847. Translated, 1950, in The Birth of Genetics. Suppl., *Genetics,* 30–32.

Vries, H. de. 1900. Sur les unités des caractères spécifiques et leur application a l'étude des hybrides. Rev. génér. botan., 12: 257–271.

Vries, H. de. 1900. Das Spaltungsgesetz der Bastarde. *Ber. deutsch. botan. Gesellsch.,* 18: 83–90.

Vries, H. de. 1901. *Die Mutationstheorie.* Vol. 1. Veit, Leipzig. 648 pp.

Warmke, H. E., and A. F. Blakeslee. 1939. Sex mechanism in polyploids of Melandrium. *Science,* 89: 391–392.

Watson, J. D., and F. H. C. Crick. 1953. Genetic implications of the structure of deoxyribonucleic acid. *Nature,* 171: 964.

Weismann, A. 1891–1892. *Essays on Heredity.* Translated by A. E. Shipley, S. Schönland, and others. Oxford University Press, 2 vols., 471; 226 pp.

Westergaard, M. 1940. Studies on cytology and sex determination in polyploid forms of *Melandrium album. Dansk. Botan. Ark.,* 10: 1–131.

Wettstein, F. von. 1925. Genetische Untersuchungen an Moosen. *Bibliogr. Genet.,* 1: 1–38.

Whiting, P. W. 1943. Multiple alleles in complementary sex determination of Habrobracon. *Genetics,* 28: 365–382.

Wilson, E. B. 1896. *The Cell in Development and Heredity.* The Macmillan Co., New York. 371 pp.; second edition, 1900, 482 pp.; third edition, 1925, 1232 pp.

Winge, Ö 1917. The chromosomes. Their number and general importance. *C. R. Lab. Carlsberg,* 13: 131–275.

Winge, Ö. 1922. One-sided masculine and sex-linked inheritance in *Lebistes reticulatus. J. Genet.,* 12: 145–162.

Winge, Ö. 1935. On haplophase and diplophase in some saccharomycetes. *C. R. Lab. Carlsberg,* 21: 71–112.

Winiwarter, H. von. 1901. Recherches sur l'ovogénèse et l'organogenèse de l'ovaire des mammifères. *Arch. Biol.,* 17: 33–199.

Wright, S. 1917. Color inheritance in mammals. *J. Hered.,* 8: 224–235.

Wright, S. 1932. The roles of mutation, inbreeding, crossbreeding, and selection in evolution. *Proc. VI Internat. Congr. Genetics,* 1: 356–366.

Yule, G. U. 1902. Mendel's laws and their probable relation to intraracial heredity. New Phytol., 1: 193–207, 222–238.

Zeleny, C. 1921. The direction and frequency of mutation in the bar-eye series of multiple allelomorphs of Drosophila. *J. Exper. Zool.,* 34: 203–233.

Zirkle, C. 1935. *The Beginnings of Plant Hybridization.* University of Pennsylvania Press, Philadelphia. 239 pp.

INDEX

REMEMBERING STURTEVANT[*]

Alfred Henry Sturtevant (1891–1970) was the youngest of six children of Alfred Henry and Harriet (Morse) Sturtevant. His grandfather, Julian Sturtevant, was a Yale graduate, a Congregational minister, and one of the founders and later president of Illinois College in Jacksonville, Illinois. Sturtevant's father taught mathematics for a while at that college, but later took up farming, first in Illinois and later in southern Alabama, where the family moved when Sturtevant was seven years old. Sturtevant went to a one-room country school and later to a public high school in Mobile.

At the age of 17, Sturtevant entered Columbia University, where his brother Edgar, who was 16 years older, was teaching at Barnard College. Edgar and his wife took the young Sturtevant into their family, and Alfred lived with them while attending the University. Edgar was a scholar who later became a professor of linguistics at Yale and an authority on the Hittite language. Sturtevant said that he learned the aims and standards of scholarship and research from Edgar. It was a great pleasure for Sturtevant when he and Edgar were awarded honorary degrees at the same Yale commencement many years later. Also present at the ceremony were Sturtevant's nephew, Julian (Edgar's son), Professor (now Emeritus) of Organic Chemistry at Yale, and Sturtevant's elder son, William, then a graduate student in Yale's department of anthropology and now curator of anthropology at the Smithsonian Institution in Washington.

Sturtevant said that he became interested in genetics as the result of tabulating the pedigrees of his father's horses. He continued this interest at Columbia and also collected data on his own pedigree. At Edgar's suggestion he went to the library and read some books on heredity, with

---

[*] Reprinted from Lewis, E. B., 1995. Remembering Sturtevant, *Genetics*, **141**:1227-1230. Used with permission.

the result that he read the textbook on Mendelism by Punnett.

Sturtevant saw at once that Mendelism could explain some of the complex patterns of inheritance of coat colors in horses that he and others before him had observed. Edgar encouraged Sturtevant to write an account of his findings and take it to Morgan, who at that time was Professor of Zoology at Columbia, and from whom Sturtevant had taken a course in zoology during his freshman year. Morgan encouraged Sturtevant to publish the paper, and it was submitted to the *Biological Bulletin* in June, 1910, at the end of his sophomore year. The paper appeared that same year (Sturtevant 1910). The connection between the genetics of horses and that of Drosophila will be familiar to readers of this column from the *Perspectives* by Snell and Reed (1993) on the mouse geneticist W. E. Castle.

The other result of Sturtevant's interest in the pedigrees of horses was that he was given a desk in the famous fly room at Columbia University where, only three months before, Morgan had found the first white-eyed fly. These stories and more about the early days at Columbia, when modern genetics was in a very real sense born, are a matter of record, especially in the writings of Sturtevant himself (1965a,b).

Sturtevant once wrote that he knew of no one else at the time who was so thoroughly committed to the experimental approach to biological problems as was Morgan. It was Morgan's aim to produce a mechanistic, as opposed to a purposive, interpretation of biological phenomena. A great deal of this approach clearly rubbed off on Sturtevant.

Sturtevant had a remarkable memory. It was as if his memory were composed of a plethora of matrices waiting to be filled with any data that lent themselves to classification into discrete categories. The data might be in the form of numbers and kinds of bristles missing in a mutant fly; numbers of snails with a right-handed coil vs. a left-handed coil, the genetics of which Sturtevant was the first to explain; the relation between inversion sequences in different species; or the host of other characteristics he investigated not only in Drosophila, but in irises, evening primroses, snails, moths, and many other creatures, including human beings. Whatever form the data took, the observations fell into the appropriate matrix in his memory, from which they were readily retrievable to a degree that was truly phenomenal. Sturtevant liked to refer to this as the "blockhead" approach.

The Caltech period was a time of collaboration, especially with Sterling Emerson, Theodosius Dobzhansky, George Beadle, and Jack Schultz. It was Sturtevant's style, at least after he came to Caltech in 1928 with Morgan and Bridges, to spend his mornings doing experiments. Afternoons were spent in the biology library checking on any incoming journals, few of which in any phase of biology he did not at least dip into. The pace of science was not so frenetic as it is nowadays, so there was time for extended afternoon tea sessions at which Sturtevant might bring up a paper he had read that afternoon and that had attracted his attention. These sessions were very stimulating for the graduate students in genetics and embryology who usually attended them; among the faculty in genetics, Schultz, Emerson, and Dobzhansky were likely to be present in addition to Sturtevant, and in embryology, Albert Tyler, who was working on the biochemistry of fertilization. Although a rift had developed between Sturtevant and Dobzhansky, there was no sign of it in front of the graduate students.

Sturtevant taught the undergraduate course in genetics at Caltech for many years. From time to time he also gave a course for undergraduates in entomology, complete with a field laboratory session. His lectures on topics in advanced genetics were scholarly reviews of special areas of genetics, often dealing with organisms with bizarre genetics, such as the protozoa. His lectures were especially valuable because he covered areas of research not ongoing at Caltech. The elementary course in genetics that Sturtevant taught was based on a textbook that he and George Beadle wrote (1939). It was not so widely used as perhaps it should have been, probably because it was considered too difficult for the average student. It was tailored for Caltech students, and the problems especially were a challenge, even for Caltech undergraduates.

Sturtevant and Beadle planned to revise the textbook, but the pressure of other work and the rapidity of developments that followed the discovery of the role of DNA prevented the revision. Sturtevant also liked to point out that both he and Beadle found after writing the book that each had used the term "gene" differently. For example, the white gene to Sturtevant was the specific white mutant, but to Beadle it represented the constellation of white alleles including the wild-type allele. Sturtevant facetiously blamed their inability to get out a second edition on this difference in thinking about the gene. Characteristically, he would ask each geneticist whom he met how he or she used the term, and he then promptly catalogued such persons according to whether they thought of the gene the way he did or the way Beadle did. The person asked did not, of course, need to worry about his answer being in good company in either case.

Sturtevant read widely and kept abreast of many topics of current interest, especially politics. He would, for example, read the Sunday *New York Times* and the *Manchester Guardian Weekly* virtually from cover to cover. He was especially happy if he could do the crossword puzzle in the *Guardian* at one sitting. Those who know those puzzles will understand that only a very special breed of person attempts them, let alone solves them in one sitting. In the evening he would browse through the *Encyclopedia Britannica*, which was shelved next to his easy chair. He complained one time, and he was not bragging, that he had difficulty in finding an article which he had not already read.

Sturtevant was fascinated with puzzles of all kinds, especially puzzles involving three-dimensional objects. When Anne Roe (1953) made a study of what makes scientists tick, she chose Sturtevant as one of her subjects. He was not only flattered, but overjoyed at the opportunity to take the tests, which he viewed simply as a new set of puzzles to work out.

Sturtevant would develop a topic logically and succinctly, whether he was publishing a paper or giving a formal lecture. In private conversation, however, he always seemed to assume that the listener was at least as well versed in the subject as he was, so he would leave out the preliminaries and get right to the point. This could be mystifying to some. For others it was a challenge to try to become sufficiently versed to profit by listening to his ideas or tapping the tremendous store of information at his fingertips on almost any topic of substance. His papers were so well written that one would assume that he had labored over each word. His penciled manuscripts rarely contained more than a few minor changes inserted into the original draft, which was done in longhand on foolscap. When asked how he did this, he told me that he usually spent many days mulling the paper over in his mind until all the words fell into place, and then all he had to do was write it down from memory.

Sturtevant developed a keen interest in the history of science; his book, *A History of Genetics* (1965a), is a classic. His main purpose in writing it, I believe, was to give credit where he thought it was due, always a difficult task, and at the same time to trace the history of the ideas underlying scientific discoveries. I believe he would have decried a tendency in some quarters to relate scientific discoveries to the sociopolitical views of the discoverers themselves. His fascination with pedigrees, including his own, led him to compile an appendix that contained a series of "intellectual" pedigrees. Sturtevant, of course, was a direct descendant of T. H. Morgan and of E. B. Wilson, another eminent biologist who was a contemporary and friend of Morgan's at Columbia. Morgan and Wilson were, in turn, direct descendants of Martin and Brooks, two men who were at Johns Hopkins University where Morgan had obtained his doctorate; Martin was descended from T. H. Huxley and

Brooks from Louis Agassiz; and so it went.

Sturtevant had a fund of aphorisms and anecdotes that he liked to spring whenever an occasion arose. Three of his favorites were from Morgan: "Establish a point and publish it;" or, when trying to overcome the difficulty in starting to write a paper, "Compose a flowery introduction, then throw it away and write the paper;" or, when a Drosophila experiment gave a totally unexpected result, "They will fool you every time." Sturtevant had one that pertained to his own marriage to Phoebe Reed Sturtevant and to that of a number of their friends, namely, "Marriages are made in heaven but there is a branch office in Woods Hole." A few were deliberately outrageous in order to make a subtle point: "Too bad graduate students are people;" or "Vertebrates are a mistake and should never have been invented." He liked to deflate pomposity whenever he ran across it and referred to pompous persons as "stuffed shirts." Echoing his contempt for profundity, he would say, "Something is profound if it reaches conclusions which I like by methods I don't understand."

Sturtevant's love for all living things, including people, was expressed in many ways. For example, in 1954 he gave the presidential address before the Pacific Division of the American Association for the Advancement of Science, where he warned of the potential hazards to human beings of the fallout from the atmospheric testing of atomic bombs. What had provoked Sturtevant was a strong statement issued by the executive branch of the government that the fallout levels from testing were far below any that could cause damage to human beings. This assumption, that there is a threshold for damage from ionizing radiation, had no evidence to support it and clearly was being used to justify testing of nuclear weapons.

Although I know that some assumed that the only purpose of Sturtevant's remarks was a desire to see a halt to bomb testing, this was not the case. He took a neutral stance and, although he felt there might be a need for testing, the public should be given the best estimate that scientists could make about the nature of the danger to the unborn from fallout levels of radiation. In "Quarreling Geneticists and a Diplomat," Crow (1995) has described in more detail the ways in which Sturtevant and other geneticists interacted in assessing radiation risks to the germ plasm.

I am indebted to Sturtevant's son, William, for pointing out in a personal communication that his father "had deep disdain for eugenics and a strong contempt for all forms of social discrimination," sentiments that perfectly sum up Sturtevant's position on these matters. Indeed, most of the chapter on the "Genetics of Man" in Sturtevant's *History of Genetics*

(1965a) is devoted to a balanced treatment of the nature-nurture question.

Sturtevant's scientific accomplishments have been reviewed elsewhere, by himself (1965a); by Sterling Emerson (1971), who first became acquainted with him in 1922; by G. W. Beadle (1970), who first came to Caltech in 1931 as a National Research Council Fellow; and by me (1976). Some of his most important papers were reprinted in a book, *Genetics and Evolution* (1961), on the occasion of his 70th birthday. Sturtevant was invited to make addenda to those papers as he saw fit; characteristically, he made only the briefest possible ones.

In a *Perspectives,* J. F. Crow (1988) stressed Sturtevant's remarkable contributions to virtually every branch of genetics. One of Sturtevant's most enduring scientific interests was that of evolutionary theory and how to approach it experimentally. One of his first contributions relevant thereto was his discovery and analysis of hybrids between *Drosophila melanogaster* and *D. simulans,* for which there is a valuable *Perspectives* by W. F. Provine (1991).

Sturtevant's research style was to let the experiments lead the way. In this respect he was not restrained by having to write grant proposals, and a decline in his rate of publishing after 1945 might have resulted in a low score anyway. Bateson is often cited for having said, "Treasure your exceptions." I believe Sturtevant's admonition would be, "Analyze your exceptions," for it is his remarkable analytical ability that shines through all his work.

For Sturtevant, science must have been an exciting and rewarding journey into the unknown. It was fortunately a long journey, with detours to many realms, and I am sure he savored every minute of it.

E. B. Lewis
Biology Division
California Institute of Technology
Pasadena, California

LITERATURE CITED

Beadle, G. W., 1970 Alfred Henry Sturtevant (1891–1970), *Year Book of The American Philosophical Society*, pp. 166–171.

Crow, J. F., 1988 A diamond anniversary: the first chromosome map. *Genetics* 118: 1–3.

Crow, J. F., 1995 Quarreling geneticists and a diplomat. *Genetics* 140: 421–426.

Emerson, S., 1971 Alfred Henry Sturtevant (November 21, 1891–April 6, 1970). *Annu. Rev. Genet.* 5: 1–4.

Lewis, E. B., 1976 Sturtevant, Alfred Henry, pp. 133–138 in *Dictionary of Scientific Biography,* vol. 13, edited by C. C. Gillispie. Charles Scribner's Sons, New York.

Provine, W. B., 1991 Alfred Henry Sturtevant and crosses between *Drosophila melanogaster* and *Drosophila simulans. Genetics* 129: 1–5.

Roe, A., 1953 *The Making of a Scientist.* Dodd, Mead & Co., New York.

Snell, G. D., and S. Reed, 1993 William Ernest Castle, pioneer mammalian geneticist. *Genetics* 133: 751–753.

Sturtevant, A. H., 1910 On the inheritance of color in the American harness horse. *Biol. Bull.* 19: 204–216.

Sturtevant, A. H., 1961 *Genetics and Evolution: Selected Papers of A. H. Sturtevant,* edited by E. B. Lewis with a foreword by G. W. Beadle. W. H. Freeman, San Francisco.

Sturtevant, A. H., 1965a *A History of Genetics.* Harper & Row, New York.

Sturtevant, A. H., 1965b The "fly room." *Am. Sci.* 53: 303–307.

Sturtevant, A. H., and G. W. Beadle, 1939 *An Introduction to Genetics.* W. B. Saunders Co., Philadelphia.